Building Mathematics Skills

New Century Edition

William F. Hunter
formerly Clinical Psychologist
Minneapolis, Minnesota

Pauline L. LaFollette
formerly Teacher
Fort Wayne Community School
Fort Wayne, Indiana

Greta Kae Smith
Coordinator of Special Education
Fort Plain Central School
Fort Plain, New York

The Learning Skills Series: Mathematics
Acquiring Mathematics Skills
Building Mathematics Skills
Continuing Mathematics Skills
Directing Mathematics Skills

ISBN 0-7915-3179-1 (formerly The Learning Skills Series: Arithmetic)

Phoenix Learning Resources, Inc.
New York

2 3 4 5 6 7 8 9 10 09 08 07 06 05

Contents

*[a] indicates assessment

What do you think this man is doing ?

How is he doing it ?

Why does he do it this way ?

Is this man counting or adding ?

Use sticks to count the people in this room.

This is another way to count. This is a tally stick.

How do you think it works ?

Make a tally stick ?

Use it to count the people who live at your house.

What are some other ways to count ?

Finish this counting chart.

1	2	3							10
				15			18		
								29	
				56					
		74							
					97				

Count by 10s.

_____ _____ _____ _____ _____ *60* _____ _____ _____ _____

Count by 5s.

_____ _____ _____ _____ _____ _____ *35* _____ _____ _____

_____ _____ _____ *70* _____ _____ _____ _____ _____ _____

Count by 2s.

2 _____ _____ _____ _____ _____ *14* _____ _____ _____

_____ _____ _____ _____ _____ _____ _____ *36* _____ _____

_____ _____ *46* _____ _____ _____ _____ _____ *58* _____

_____ *64* _____ _____ _____ *74* _____ _____ _____ _____

_____ _____ *88* _____ _____ _____ *96* _____ _____ _____

2

Add.

1. 24
   + 7

2. 33
   + 19

3. 49
   + 2

4. 58
   + 3

5. 97
   + 5

6. 85
   + 8

7. 96
   + 6

8. 89
   + 4

9. 77
   + 13

10. 68
    + 9

11. 87
    + 6

12. 96
    + 8

13. 85
    + 7

14. 86
    + 9

15. 73
    + 36

16. 74
    + 56

17. 65
    + 46

18. 80
    + 57

19. 31
    + 27

20. 72
    + 27

21. $4000 + 200 + 60 + 7 =$ _____

22. $2000 + 300 + 50 + 8 =$ _____

23. $1000 + 40 + 8 =$ _____

24. $2042 =$ _____ $+$ _____ $+$ _____

25. $7241 =$ _____ $+$ _____ $+$ _____ $+$ _____

26. Three thousand six hundred forty-two = _____

27. Five thousand one hundred thirty-three = _____

28. Six thousand five hundred ninety-five = _____

29. Two thousand six hundred six = _____

30. One thousand fourteen = _____

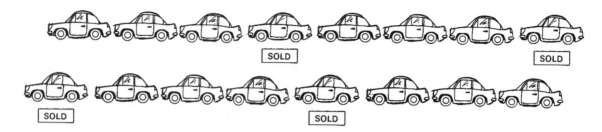

There were 25 cars for sale.
13 were sold.
How many are left ? _____

There were 73 cars on another lot for sale.
27 were sold.
How many are left ?_____

Subtract.

1. 89 − 63	**2.** 48 − 24	**3.** 67 − 26	**4.** 82 − 11	**5.** 76 − 23
6. 95 − 82	**7.** 37 − 26	**8.** 44 − 12	**9.** 78 − 15	**10.** 62 − 10
11. 92 − 86	**12.** 73 − 25	**13.** 63 − 15	**14.** 83 − 27	**15.** 64 − 15

Write the numerals.

16. Four thousand, three hundred, sixty seven = _____

17. One thousand, twenty four = _____

18. Two thousand, seven hundred, one = _____

19. Eight thousand, five hundred, seventy six = _____

20. Two thousand, four hundred fourteen = _____

Subtract.

1. 34
    − 5

2. 41
    − 3

3. 60
    − 1

4. 24
    − 8

5. 19
    − 5

6. 20
    − 6

7. 12
    − 3

8. 21
    − 3

9. 40
    − 2

10. 67
    − 8

11. 96
    − 9

12. 50
    − 7

13. 41
    − 4

14. 11
    − 2

15. 60
    − 4

16. 75
    − 6

17. 34
    − 9

18. 70
    − 8

19. 12
    − 9

20. 23
    − 6

Write your own subtraction word problem. Then solve it.

Mary picked 30 apples.
Tom picked 23 apples.
Sam picked 18 apples.
Together they picked _____ apples.

Yesterday Tom picked 42 apples.
Mom used 14 of them to bake a pie.
How many apples does Tom have now ? _____

Add.

1. 29 17 + 38	**2.** 26 19 + 45	**3.** 27 43 + 74	**4.** 64 55 + 48	**5.** 63 25 + 39
6. 47 25 + 30	**7.** 29 18 + 36	**8.** 86 54 + 89	**9.** 48 43 + 63	**10.** 64 15 + 23

Subtract.

11. 96 − 9	**12.** 50 − 7	**13.** 41 − 4	**14.** 11 − 2	**15.** 60 − 4
16. 75 − 6	**17.** 34 − 9	**18.** 70 − 8	**19.** 12 − 9	**20.** 23 − 6
21. 15 − 7	**22.** 34 − 8	**23.** 35 − 7	**24.** 26 − 9	**25.** 13 − 6

You have 58¢. You spend 49¢ on a
soda. How much do you have left ? _____

You have 37¢. You return
some bottles and get 55¢.
How much do you have now ? _____

Subtract.

1. 60 − 4	**2.** 97 − 9	**3.** 75 − 3	**4.** 93 − 8	**5.** 47 − 8

6. 64 − 6	**7.** 96 − 7	**8.** 12 − 9	**9.** 40 − 5	**10.** 34 − 14

Add.

11. 48 + 27	**12.** 82 + 48	**13.** 67 + 27	**14.** 84 + 15	**15.** 62 + 83

16. 721 + 362	**17.** 809 + 147	**18.** 638 + 246	**19.** 608 + 283	**20.** 746 + 193

21. 72 46 + 83	**22.** 83 47 + 82	**23.** 41 82 + 55	**24.** 62 38 + 49	**25.** 70 43 + 27

Tom picked 50 apples.
Nell picked 80 apples.

Tom and Nell picked _____
apples together.

Nell picked _____
more apples than Tom.

Subtract.

1.	56 − 49	**2.**	76 − 28	**3.**	87 − 17	**4.**	97 − 57	**5.**	86 − 38
6.	63 − 17	**7.**	75 − 69	**8.**	74 − 47	**9.**	94 − 26	**10.**	84 − 48

Add.

11.	307 + 436	**12.**	310 + 872	**13.**	809 + 707	**14.**	400 + 327	**15.**	209 + 790
16.	347 + 609	**17.**	392 + 906	**18.**	283 + 608	**19.**	409 + 390	**20.**	642 + 290

Elaine has a collection
of 73 baseball cards.
She buys 20 more. How
many does she have now ? _____

Bob had 80 football cards.
He gave away 16 of them.
How many does he have now ? _____

Add.

1.   72
    + 46

2.   70
    + 38

3.   81
    + 54

4.   60
    + 15

5.   51
    + 39

6.   62
    + 29

7.   73
    + 35

8.   62
    + 47

9.   80
    + 53

10.   92
    + 46

Subtract.

11.   51
    − 18

12.   40
    − 17

13.   93
    − 55

14.   50
    − 34

15. 119
    − 30

16.   48
    − 42

17.   59
    − 35

18.   57
    − 19

19.   51
    − 34

20.   85
    − 63

21.   74
    − 40

22.   92
    − 89

23.   80
    − 78

24.   32
    − 27

25.   90
    − 57

26.   21
    − 18

27.   53
    − 47

28.   61
    − 35

29.   71
    − 69

30.   82
    − 75

Count by 2s.

2 ___ ___ ___ ___ 12 ___ ___ ___ ___

___ 24 ___ ___ ___ ___ 36 ___ ___

___ ___ 48 ___ ___ ___ ___ ___ 60

1.	39 24 + 44	2.	31 26 + 23	3.	18 25 38 + 49	4.	27 53 49 + 57	5.	45 63 36 + 28

6.	18 76 27 + 99	7.	65 14 34 + 24	8.	18 28 48 + 26	9.	46 56 17 + 47	10.	87 55 67 + 74

11.	75 19 57 + 39	12.	59 26 97 + 46	13.	79 47 63 + 91	14.	81 35 68 + 92	15.	18 27 45 + 18

16.	17 28 44 + 19	17.	80 36 67 + 92	18.	78 46 64 + 93	19.	76 20 55 + 37	20.	58 27 96 + 47

Count by 2s.

_2_ ___ ___ ___ ___ ___ ___ ___ ___ _20_

_30_ ___ ___ ___ ___ ___ ___ ___ _50_

_10_ ___ ___ ___ ___ ___ ___ ___ _30_

Add.

1. 78 + 14	**2.** 68 + 25	**3.** 84 + 44	**4.** 37 + 20	**5.** 53 + 33
6. 86 + 56	**7.** 49 + 26	**8.** 85 + 15	**9.** 98 + 70	**10.** 72 + 52

Subtract.

11. 96 − 94	**12.** 38 − 26	**13.** 55 − 40	**14.** 26 − 12	**15.** 54 − 38
16. 79 − 69	**17.** 90 − 80	**18.** 98 − 64	**19.** 86 − 44	**20.** 77 − 53

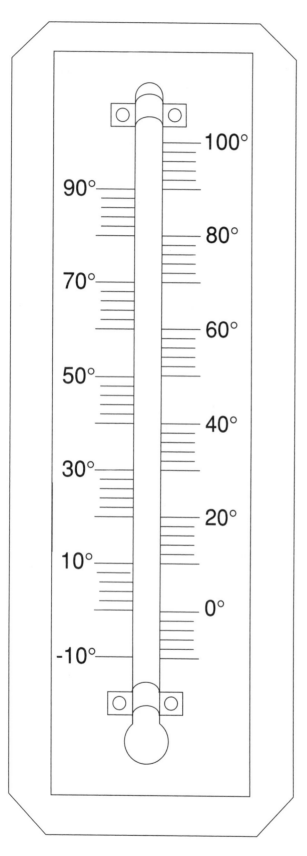

This is a picture of a

_____ .

It is used to measure

_____ .

We measure temperature in

_____ .

We might write the temperature as 45°.
What does the little circle mean ?

_____ .

Look at the thermometer. How many
degrees are there between each two
numbers ?

_____ .

Look at the spaces between 0° and 10°.
How many spaces are there ?

_____ .

Count the spaces by 2's.
Do you count to 10 ? _____

How many degrees is each little space ?

_____ .

Write the degrees on the marks between 0°
and 10°.

Write the degrees on the marks between 0°
and −10°.

What is different about these numbers ?

_____ .

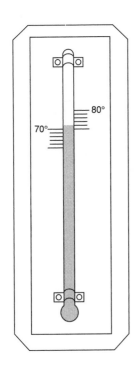

Count by 2s.

					10
20					30
30					40
50					60
70					

The temperature on
this thermometer is

_____ .

Subtract.

1.	64 − 28	**2.**	84 − 16	**3.**	77 − 50	**4.**	80 − 49	**5.**	43 − 25
6.	84 − 27	**7.**	30 − 8	**8.**	92 − 65	**9.**	83 − 51	**10.**	68 − 24
11.	91 − 49	**12.**	67 − 47	**13.**	50 − 25	**14.**	477 − 244	**15.**	309 − 105

Count by 2s. _40_ ___ ___ ___ ___ _50_

Add.

1. 98
  + 65

2. 89
  + 47

3. 78
  + 60

4. 69
  + 67

5. 89
  + 83

6. 97
  + 97

7. 24
  + 88

8. 59
  + 63

9. 24
  + 97

10. 59
  + 38

11. 76
  + 94

12. 67
  + 80

13. 125
  + 35

14. 116
  + 60

15. 105
  + 60

16. 294
  + 67

17. 590
  + 77

18. 385
  + 257

19. 296
  + 584

20. 750
  + 648

21. 382
  + 772

22. 820
  + 276

23. 634
  + 265

24. 458
  + 563

25. 575
  + 309

26. 255
  + 776

27. 367
  + 730

28. 74
   95
  + 23

29. 43
   67
  + 50

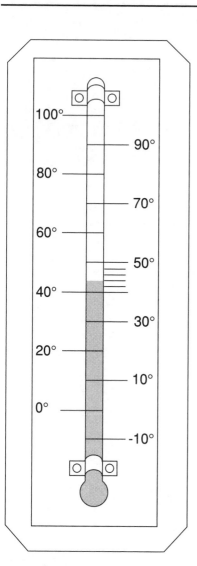

What is the temperature on this thermometer? _____

calendar

Make a calendar for October.

OCTOBER						
Sun.	Mon.	Tue.	Wed.	Thur.	Fri.	Sat.

Subtract.

1. 95
 − 68

2. 80
 − 67

3. 83
 − 39

4. 97
 − 97

5. 67
 − 59

6. 64
 − 0

7. 78
 − 54

8. 50
 − 37

9. 69
 − 40

10. 93
 − 83

11. 92
 − 67

12. 81
 − 48

13. 95
 − 23

14. 125
 − 35

15. 15
 − 10

16. 35
 − 28

17. 60
 − 40

18. 67
 − 59

Harvest Festival Sale

CARVE YOUR OWN

PUMPKINS!

Small $2.50

Medium $4.00

Large $8.00

PIES

Apple • Pumpkin • Blueberry

$3.00 each

Macintosh or Delicious

APPLES

$5.00 a bushel

Decorate for Fall

small wreath

$9.00

large wreath

$12.00

CIDER

$2.39 a quart

---

1. 2 bushels of apples would cost _____ .

2. If you had $10.00 and bought a small

   wreath, your change would be _____ .

3. What are the sizes of 2 of the pumpkins ? _____   _____

   How much would the 2 pumpkins cost? _____ .

4. An apple pie and a blueberry pie would cost _____ .

5. You had $5.00, you bought cider. Your change

   was _____ .

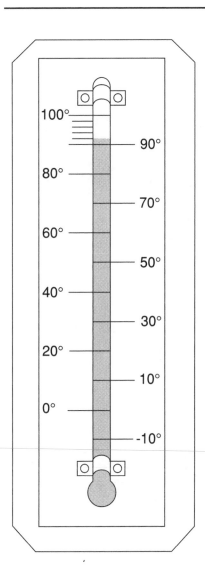

Temperature _____

Count by 2s.

0 —— —— —— —— 10

20 —— —— —— 28 ——

50 —— 54 —— 60 ——

Add.

1.	76	**2.**	43	**3.**	54	**4.**	85
	16		68		15		73
	+ 15		+ 72		+ 44		+ 58

5.	45	**6.**	27	**7.**	52	**8.**	72
	+ 65		+ 86		+ 80		+ 47

9.	39	**10.**	48	**11.**	43	**12.**	58
	23		85		34		65
	+ 15		+ 26		+ 14		+ 32

13.	24	**14.**	32	**15.**	83	**16.**	35	**17.**	61	**18.**	13
	+ 46		+ 57		+ 15		+ 27		+ 85		+ 57

19.
0	1	2	3	4	5	6	7	8	9	10
× 2	× 2	× 2	× 2	× 2	× 2	× 2	× 2	× 2	× 2	× 2

20.
43	61	32	50	84	72	93
× 2	× 2	× 2	× 2	× 2	× 2	× 2

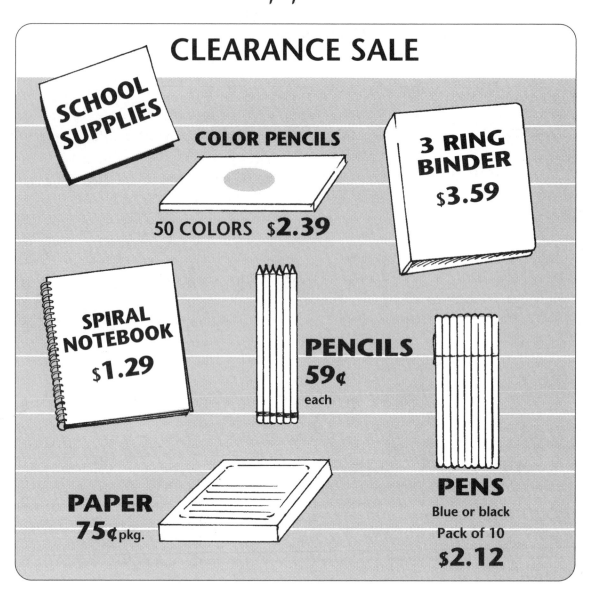

CLEARANCE SALE

SCHOOL SUPPLIES

COLOR PENCILS

50 COLORS **$2.39**

3 RING BINDER $3.59

SPIRAL NOTEBOOK $1.29

PENCILS 59¢ each

PAPER 75¢ pkg.

PENS
Blue or black
Pack of 10
$2.12

1. How much for one pencil ? _____

   How much for two pencils ? _____

2. How much for one notebook

   and one pack of pens ? _____

3. How much for two spiral notebooks ? _____

4. How much for two boxes of color pencils ? _____

5. How much for two binders and

   two packs of pens ? _____

Count by 5s.

5	—— —— —— —— —— —— —— ——	50
20	—— —— —— —— —— —— —— ——	65
40	—— —— —— —— —— —— —— ——	85
55	—— —— —— —— —— —— —— ——	100

1.

5	8	3	5	7	5	4	9	5	6
× 0	× 5	× 5	× 2	× 5	× 1	× 5	× 5	× 5	× 5

2.

42	73	64	80	12	53	92
× 5	× 5	× 5	× 5	× 5	× 5	× 5

3.

36	83	46	87	41	59	27
× 2	× 2	× 2	× 2	× 2	× 2	× 2

4.

83	76	93	70	61	25	52
× 5	× 2	× 2	× 5	× 5	× 6	× 5

FRESH FRUIT

**Great
GRAPES
$1.58 lb.**

**Sweet
PEARS
50¢ each**

**Fresh
STRAWBERRIES
$2.99 a basket**

**Florida
ORANGES
4 lb bag $1.99**

**WATERMELON
$4.00 each**

1. A bag of oranges and a pound of grapes would cost _____ .

2. You had $5.00. You bought a watermelon. How much is left ? _____

3. Five pears would cost _____ .

4. Two baskets of strawberries would cost _____ .

5. You had $3.00. You bought a pound of grapes. What is your change ? _____

Count by 2s.   _O_ __ __ __ __ __ __ __ __ __ _20_

Count by 5s.   _O_ __ __ __ __ __ __ __ __ __ _50_

Count by 10s.   _O_ __ __ __ __ __ __ __ __ __ _100_

1. 42 × 2	**2.** 53 × 5	**3.** 46 × 5	**4.** 82 × 5	**5.** 83 × 2
6. 31 × 5	**7.** 67 × 2	**8.** 97 × 5	**9.** 57 × 2	**10.** 88 − 5

Subtract.

11. 52 − 31	**12.** 33 − 12	**13.** 40 −35	**14.** 50 − 49	**15.** 62 − 54
16. 72 − 20	**17.** 28 − 13	**18.** 87 − 35	**19.** 82 − 72	**20.** 70 − 60
21. 35 − 27	**22.** 94 − 65	**23.** 46 −36	**24.** 82 − 80	**25.** 20 − 14
26. 56 − 48	**27.** 45 − 26	**28.** 53 − 48	**29.** 34 − 25	**30.** 30 − 16
31. 63 − 50	**32.** 32 − 22	**33.** 83 −15	**34.** 75 − 58	**35.** 41 − 26
36. 52 − 45	**37.** 65 − 32	**38.** 90 − 30	**39.** 34 − 17	**40.** 524 − 203

Subtract.

1.	$3.00 − $1.00 $	**2.**	$0.50 − $0.35 $	**3.**	$17.00 − $12.00 $	**4.**	$1.00 − $0.50 $

5.	$20.00 − $10.00 $	**6.**	$5.90 − $4.08 $	**7.**	60¢ − 5¢ ¢	**8.**	$10.00 − $ 8.00 $

Add.

9.	$3.00 $5.00 + $2.00 $	**10.**	$7.00 $1.00 + $2.00 $	**11.**	$4.00 $3.00 + $6.00 $	**12.**	$3.00 $7.00 + $2.00 $

13.	$0.30 $0.20 + $0.90 $	**14.**	$6.00 $1.00 + $5.00 $	**15.**	15¢ 9¢ + 5¢ ¢	**16.**	$2.00 $0.25 + $7.00 $

Multiply.

17.	73 × 5	**18.**	56 × 2	**19.**	314 × 5	**20.**	741 × 5

Divide.

21. $5\overline{)45}$ **22.** $2\overline{)18}$ **23.** $2\overline{)16}$ **24.** $5\overline{)30}$ **25.** $5\overline{)15}$

Tell the time.

_____   _____   _____   _____   _____

Count by 5s. ____ ____ ____ ____ ____

____ ____ ____ ____ ____ ____

Multiply.

1. 24	**2.** 83	**3.** 75	**4.** 18	**5.** 90
× 5	× 5	× 5	× 5	× 5

$5 \times 0 = 0$				
$5 \times 1 = 5$	**6.** 67	**7.** 10	**8.** 62	**9.** 49
$5 \times 2 = 10$	× 5	× 5	× 5	× 5
$5 \times 3 = 15$				
$5 \times 4 = 20$	**10.** 57	**11.** 98	**12.** 54	**13.** 20
$5 \times 5 = 25$	× 5	× 5	× 5	× 5
$5 \times 6 = 30$				
$5 \times 7 = 35$				
$5 \times 8 = 40$	**14.** 36	**15.** 13	**16.** 78	**17.** 96
$5 \times 9 = 45$	× 5	× 5	× 5	× 5

Divide.

18. $7\overline{)35}$ **19.** $5\overline{)40}$ **20.** $5\overline{)25}$ **21.** $3\overline{)15}$

22. $5\overline{)5}$ **23.** $5\overline{)45}$ **24.** $5\overline{)35}$ **25.** $6\overline{)30}$

Show the temperatures.

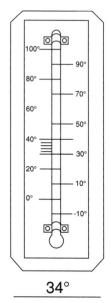
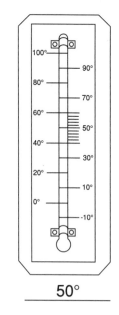
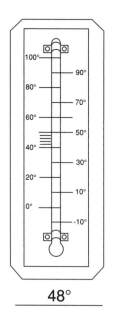
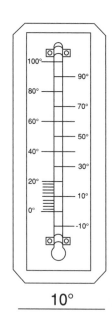

| 34° | 50° | 48° | 10° |

1.
$$\begin{array}{r} 21 \\ \times\ 2 \\ \hline \end{array}$$
$$\begin{array}{r} 73 \\ \times\ 2 \\ \hline \end{array}$$
$$\begin{array}{r} 46 \\ \times\ 2 \\ \hline \end{array}$$
$$\begin{array}{r} 50 \\ \times\ 2 \\ \hline \end{array}$$
$$\begin{array}{r} 89 \\ \times\ 2 \\ \hline \end{array}$$
$$\begin{array}{r} 35 \\ \times\ 2 \\ \hline \end{array}$$

2.
$$\begin{array}{r} 78 \\ \times\ 2 \\ \hline \end{array}$$
$$\begin{array}{r} 38 \\ \times\ 2 \\ \hline \end{array}$$
$$\begin{array}{r} 19 \\ \times\ 2 \\ \hline \end{array}$$
$$\begin{array}{r} 20 \\ \times\ 2 \\ \hline \end{array}$$
$$\begin{array}{r} 58 \\ \times\ 2 \\ \hline \end{array}$$
$$\begin{array}{r} 62 \\ \times\ 2 \\ \hline \end{array}$$

3.
$$\begin{array}{r} 94 \\ \times\ 2 \\ \hline \end{array}$$
$$\begin{array}{r} 70 \\ \times\ 2 \\ \hline \end{array}$$
$$\begin{array}{r} 13 \\ \times\ 2 \\ \hline \end{array}$$
$$\begin{array}{r} 86 \\ \times\ 2 \\ \hline \end{array}$$
$$\begin{array}{r} 49 \\ \times\ 2 \\ \hline \end{array}$$
$$\begin{array}{r} 44 \\ \times\ 2 \\ \hline \end{array}$$

4.
$$\begin{array}{r} 68 \\ \times\ 2 \\ \hline \end{array}$$
$$\begin{array}{r} 90 \\ \times\ 2 \\ \hline \end{array}$$
$$\begin{array}{r} 53 \\ \times\ 2 \\ \hline \end{array}$$
$$\begin{array}{r} 124 \\ \times\ 2 \\ \hline \end{array}$$
$$\begin{array}{r} 75 \\ \times\ 2 \\ \hline \end{array}$$
$$\begin{array}{r} 206 \\ \times\ 2 \\ \hline \end{array}$$

$2 \times 0 =$ _____

$2 \times 1 =$ _____

$2 \times 2 =$ _____

$2 \times 3 =$ _____

$2 \times 4 =$ _____

$2 \times 5 =$ _____

$2 \times 6 =$ _____

$2 \times 7 =$ _____

$2 \times 8 =$ _____

$2 \times 9 =$ _____

Subtract.

1.   43
      − 25

2.   80
      − 17

3.   58
      − 13

4.   46
      − 9

5.   62
      − 8

6.   59
      − 48

7.   48
      − 35

8.   40
      − 7

9.   39
      − 18

10.   55
       − 19

11.   34
       − 19

12.   65
       − 40

13.   90
       − 28

14.   35
       − 23

15.   20
       − 15

16.   83
       − 65

17.   28
       − 14

18.   54
       − 9

19.   65
       − 15

20.   85
       − 49

21.   278
       − 173

22.   32
       − 29

23.   57
       − 40

24.   396
       − 128

We have learned that we use numbers to count things.
When we count, we find how many.
Sometimes we add to find how many.
Sometimes we multiply to find how many.
Numbers are used in many games. You use numbers
to count when you play dominoes.
First, each player takes 5 tiles. The player with the
largest double tile plays first, like this:

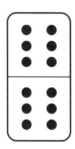

The next player must have a tile with the same number on it.
She plays like this:

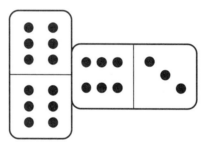

Now add the numbers on the ends. 6 + 6 + 3 = 1 5
The second player wins 15 points.
You win points only when you have 5 or 1 0 or 1 5 points.
This is counting by 5's.
The next player plays this tile:

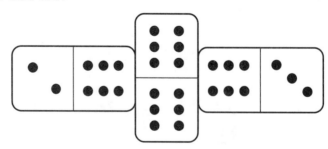

Does this player win any points ? _____

How many points does he win ? _____   2 + 3 = _____

Use dominoes to help with multiplication facts.

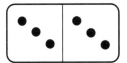 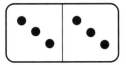 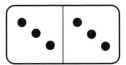 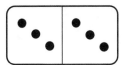 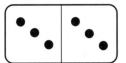

Multiply.

1.

3	3	3	3	3	3	3	3	3
× 1	× 2	× 3	× 4	× 5	× 6	× 7	× 8	× 9

Add.

2.

15	34	69	58	71
+ 75	+ 76	+ 73	+ 27	+ 77

3.

95	10	76	29	75
+ 66	+ 5	+ 43	+ 38	+ 56

4.

86	74	86	90	46
+ 29	+ 81	+ 53	+ 58	+ 54

5.

73	91	51	42	74
+ 65	+ 18	+ 64	+ 75	+ 66

Multiply.

6.

27	36	23	12	34
× 3	× 3	× 5	× 3	× 3

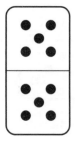
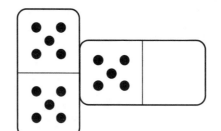

$$\begin{array}{r} 5 \\ + 5 \\ \hline \end{array} \qquad \begin{array}{r} 5 \\ \times 2 \\ \hline \end{array}$$

$5 + 5 + 5 =$ _____

$5 \times 3 =$ _____

	3 teaspoons (tsp)	= 1 tablespoon (tbs)
Study this table:	16 tablespoons (tbs)	= 1 cup (c)
	2 cups (c)	= 1 pint (pt)
	2 pints (pt)	= 1 quart (qt)

Subtract.

1. $\begin{array}{r} 14 \\ -\ 8 \\ \hline \end{array}$	$\begin{array}{r} 25 \\ -\ 9 \\ \hline \end{array}$	$\begin{array}{r} 48 \\ -20 \\ \hline \end{array}$	$\begin{array}{r} 33 \\ -26 \\ \hline \end{array}$	$\begin{array}{r} 70 \\ -25 \\ \hline \end{array}$
2. $\begin{array}{r} 54 \\ -38 \\ \hline \end{array}$	$\begin{array}{r} 23 \\ -17 \\ \hline \end{array}$	$\begin{array}{r} 41 \\ -26 \\ \hline \end{array}$	$\begin{array}{r} 54 \\ -49 \\ \hline \end{array}$	$\begin{array}{r} 85 \\ -27 \\ \hline \end{array}$
3. $\begin{array}{r} 95 \\ -45 \\ \hline \end{array}$	$\begin{array}{r} 30 \\ -18 \\ \hline \end{array}$	$\begin{array}{r} 22 \\ -13 \\ \hline \end{array}$	$\begin{array}{r} 59 \\ -30 \\ \hline \end{array}$	$\begin{array}{r} 72 \\ -19 \\ \hline \end{array}$
4. $\begin{array}{r} 60 \\ -34 \\ \hline \end{array}$	$\begin{array}{r} 43 \\ -35 \\ \hline \end{array}$	$\begin{array}{r} 84 \\ -28 \\ \hline \end{array}$	$\begin{array}{r} 76 \\ -36 \\ \hline \end{array}$	$\begin{array}{r} 56 \\ -38 \\ \hline \end{array}$
5. $\begin{array}{r} 63 \\ -40 \\ \hline \end{array}$	$\begin{array}{r} 90 \\ -54 \\ \hline \end{array}$	$\begin{array}{r} 25 \\ -19 \\ \hline \end{array}$	$\begin{array}{r} 59 \\ -27 \\ \hline \end{array}$	$\begin{array}{r} 70 \\ -63 \\ \hline \end{array}$

Multiply.

6. $\begin{array}{r} 32 \\ \times\ 3 \\ \hline \end{array}$	$\begin{array}{r} 41 \\ \times\ 3 \\ \hline \end{array}$	$\begin{array}{r} 70 \\ \times\ 3 \\ \hline \end{array}$	$\begin{array}{r} 81 \\ \times\ 3 \\ \hline \end{array}$	$\begin{array}{r} 92 \\ \times\ 3 \\ \hline \end{array}$

1 pint　　　　=　_____ cups

2 pints　　　=　_____ quart

3 teaspoons　=　_____ tablespoon

1 tablespoon　=　_____ teaspoons

1 cup　　　　=　_____ tablespoons

1 quart　　　=　_____ pints

Try to find the number of dots on each tile without counting.

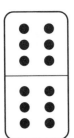 ...	6 + 6

```
         6
       + 6
       ____


         5
       + 5
       ____


         4
       + 4
       ____


         3
       + 3
       ____


         2
       + 2
       ____


         1
       + 1
       ____


         0
       + 0
       ____


         6
       + 4
       ____


         5
       + 0
       ____


         2
       + 3
       ____


         4
       + 1
       ____


         6
       + 3
       ____
```

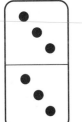
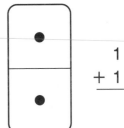
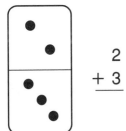

Try to add without counting.

1.	63	46	57	15	32
	+ 62	+ 14	+ 3	+ 10	+ 32

2.	40	24	606	436	517
	+ 60	+ 24	+ 606	+ 124	+ 103

3.	534	204	636	647	551
	+ 306	+ 204	+ 206	+ 143	+ 504

4.	234	402	654	663	155
	+ 230	+ 402	+ 601	+ 642	+ 100

5.	743	646	363	465	715
	+ 363	+ 406	+ 262	+ 140	+ 310

6.	324	240	460	643	651
	+ 326	+ 240	+ 460	+ 612	+ 401

7.	753	420	463	664	755
	+ 303	+ 620	+ 402	+ 614	+ 305

8.	123	404	265	466	637
	+ 423	+ 604	+ 260	+ 164	+ 623

What time is it ? _____

What time is it ? _____

Add the two ends of these dominoes.

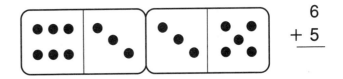

$$\begin{array}{r} 6 \\ + 5 \\ \hline \end{array}$$

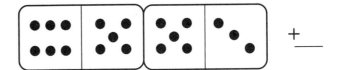

$$\begin{array}{r} \\ + \\ \hline \end{array}$$

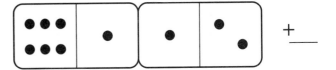

$$\begin{array}{r} \\ + \\ \hline \end{array}$$

Write the multiplication facts for 2.

Write the multiplication facts for 5.

Count by 3s to 99.

_21_ ___ ___   ___ ___ ___ ___ ___ ___

___ ___ _57_ ___ ___ ___ ___ ___ ___

___ ___ ___ ___ ___ ___

0	1	2	3	4	5	6	7	8	9
× 3	× 3	× 3	× 3	× 3	× 3	× 3	× 3	× 3	× 3

____ 🥄 = 1 🥄   ____ 🥄 = 1 🥛   ____ 🥛 = 1

tsp   tbsp   tbsp   c   c   pt

Multiply.

1.

5	4	6	3	2	4	6	8	9
× 3	× 2	× 1	× 3	× 2	× 3	× 2	× 3	× 1

2.

5	2	7	5	4	6	8	7	6
× 5	× 3	× 2	× 2	× 5	× 3	× 2	× 3	× 5

Add without counting.

3.

6	3	6	2	7	5	2	3	6
+ 5	+ 6	+ 6	+ 6	+ 3	+ 5	+ 8	+ 2	+ 4

4.

2	3	9	0	8	6	4	4	9
+ 2	+ 3	+ 1	+ 7	+ 2	+ 3	+ 6	+ 4	+ 0

Add.

5.

62	36	63	22	75	36
+ 52	+ 64	+ 62	+ 68	+ 35	+ 32

6.

99	40	54	86	63	39
+ 10	+ 74	+ 66	+ 23	+ 33	+ 61

7.

60	25	78	56	24	34
+ 67	+ 66	+ 32	+ 53	+ 86	+ 25

8.

69	26	66	32	72	53
+ 40	+ 22	+ 56	+ 66	+ 38	+ 52

9.

26	32	30	95	89	64
+ 84	+ 22	+ 37	+ 16	+ 20	+ 34

Multiply.

1. 42 × 3	87 × 1	62 × 3	70 × 3	43 × 3	52 × 3
2. 51 × 5	84 × 2	63 × 3	90 × 5	83 × 3	72 × 3

Subtract.

3. 18 − 9	25 − 17	42 − 26	32 − 14	70 − 15	54 − 29
4. 80 − 30	41 − 12	14 − 7	93 − 76	20 − 13	72 − 54
5. 41 − 27	60 − 45	43 − 15	90 − 40	61 − 53	70 − 9
6. 83 − 54	37 − 29	68 − 39	96 − 29	54 − 18	64 − 27
7. 86 − 17	91 − 32	62 − 54	85 − 38	74 − 29	43 − 15
8. 84 − 26	56 − 48	53 − 47	61 − 44	52 − 23	41 − 19
9. 70 − 35	44 − 26	50 − 40	67 − 20	22 − 18	30 − 25

BURGER WORLD
Grand Opening Menu

Hamburger
$ .79

Cheeseburger
$ .94

Chicken Sandwich
$ 2.99

World Famous Burger
with lettuce and tomato
$ 1.98

Root Beer
small **$ .85**
large **$ 1.23**

Fries
$ 1.09

Onion Rings
$ .99

Ice Cream Cone
$ .59

Apple Pie
$ .79

34

1. How much would 3
   large root beers cost ? _____
2. How much would 5
   cheeseburgers cost ? _____
3. How much would 2
   World Famous Burgers cost ? _____
4. How much would 3
   apple pies cost ? _____
5. Three chicken
   sandwiches cost _____.

Multiply.

6.	47 $\times\ 3$	82 $\times\ 3$	79 $\times\ 3$	64 $\times\ 3$	70 $\times\ 3$

7.	123 $\times\ \ \ 3$	245 $\times\ \ \ 5$	632 $\times\ \ \ 2$	741 $\times\ \ \ 3$	809 $\times\ \ \ 3$

Add.

8.	59 $+\ 10$	64 $+\ 36$	96 $+\ 14$	53 $+\ 47$	32 $+\ 17$

9.	62 $+\ 80$	85 $+\ 26$	52 $+\ 68$	78 $+\ 20$	47 $+\ 77$

10.	43 $+\ 48$	95 $+\ 23$	63 $+\ 43$	59 $+\ 45$	62 $+\ 72$

1. How would you count four nickels ? ___5¢___   _____   _____   _____

2. Another way to count four nickels is to say 4 × 5¢ = _____ ¢.

3. Then three nickels equals ___3___ × 5¢, or _____ ¢.

4. Five dimes equals _____ × 10¢, or _____ ¢.

5. Three quarters equals _____ × 25¢, or _____ ¢.

Multiply or add.

6.
15¢	40¢	35¢	25¢	10¢
× 2¢	× 5¢	× 5¢	× 4¢	× 3¢
¢	¢	¢	¢	¢

7.
85¢	60¢	90¢	90¢	75¢
× 5¢	× 2¢	× 2¢	+ 90¢	× 3¢
¢	¢	¢	¢	¢

8.
63¢	27¢	81¢	45¢	45¢
× 3¢	× 5¢	× 5¢	× 2¢	+ 45¢
¢	¢	¢	¢	¢

9.
30¢	49¢	65¢	65¢	29¢
× 3¢	× 5¢	× 2¢	+ 65¢	× 5¢
¢	¢	¢	¢	¢

10.
80¢	20¢	50¢	50¢	70¢
× 3¢	× 5¢	× 2¢	+ 50¢	× 3¢
¢	¢	¢	¢	¢

11.
95¢	45¢	45¢	15¢	15¢
× 5¢	× 3¢	45¢	× 3¢	15¢
¢	¢	+45¢	¢	+ 15¢
		¢		¢

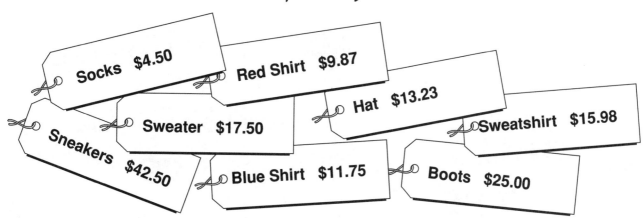

Three pairs of socks would cost _____ .

Subtract.

1. 73 − 66	96 − 34	75 − 14	58 − 47	86 − 57	78 − 69
2. 83 −47	61 − 32	29 − 11	81 − 72	24 − 15	72 − 58
3. 20 − 8	46 − 29	98 − 68	83 − 74	43 − 32	75 − 49
4. 65 −17	70 − 52	72 − 47	76 − 25	68 − 64	83 − 40
5. 68 −43	86 − 37	60 − 27	70 − 40	91 − 60	13 − 7

Multiply.

6. 56 × 3	68 × 2	39 × 4	93 × 5	18 × 2	87 × 5
7. 45 × 2	57 × 3	36 × 5	89 × 3	54 × 5	37 × 2

SPORTING GOODS

TENNIS BALLS
$2.99

BASEBALL
$5.99

BASKETBALL
$34.99

SOFTBALLS
$9.98

GOLF BALLS
$21.99

FOOTBALL
$29.99

SOCCER BALL
$37.50

1. How many tennis balls in 1 tube ? _____
2. How many tennis balls in 3 tubes ? _____
3. How many tennis balls in 5 tubes ? _____
4. What is the cost of 3 baseballs ? _____
5. How many golf balls in 3 packages ? _____
6. What is the cost of 3 packages of golf balls ? _____
7. How much would 2 basketballs cost ? _____

Multiply.

8. $5 \times 2 =$ ____ $0 \times 3 =$ ____ $3 \times 4 =$ ____ $3 \times 5 =$ ____

9. $4 \times 5 =$ ____ $3 \times 9 =$ ____ $7 \times 5 =$ ____ $6 \times 5 =$ ____

10. $8 \times 3 =$ ____ $1 \times 5 =$ ____ $5 \times 5 =$ ____ $7 \times 3 =$ ____

11. $3 \times 2 =$ ____ $8 \times 5 =$ ____ $6 \times 3 =$ ____ $2 \times 6 =$ ____

12. $2 \times 7 =$ ____ $2 \times 8 =$ ____ $10 \times 3 =$ ____ $10 \times 5 =$ ____

Divide.

13. ____ $\times 2 = 14$ ____ $\times 3 = 18$ ____ $\times 5 = 35$

    $14 \div 2 =$ ____ $18 \div 3 =$ ____ $35 \div 5 =$ ____

14. ____ $\times 2 = 6$ ____ $\times 3 = 9$ ____ $\times 5 = 30$

    $6 \div 2 =$ ____ $9 \div 3 =$ ____ $30 \div 5 =$ ____

16. ____ $\times 5 = 20$ ____ $\times 4 = 12$ ____ $\times 5 = 15$

    $20 \div 5 =$ ____ $12 \div 4 =$ ____ $15 \div 5 =$ ____

Show the time.　　　　　　　Tell the time.

　4:10　　　　　10:35　　　_____　　　_____　　　_____

1. 1 hour = _____ minutes　　　　2. 60 minutes = _____ hour

3. 1 day = _____ hours　　　　　4. 24 hours = _____ day

5. 1 week = _____ days　　　　　6. 14 days = _____ weeks

7. 1 year = _____ months　　　　8. 12 months = _____ year

Add across. Then add down.

68	28	87	89	60	
64	97	59	97	42	
39	84	59	65	67	
86	59	7	86	85	
56	46	68	59	93	

1 tablespoon = _____ teaspoons　　　　1 pint = _____ cups

2 tablespoons = _____ teaspoons　　　　2 pints = _____ cups

5 tablespoons = _____ teaspoons　　　　4 pints = _____ cups

3 tablespoons = _____ teaspoons　　　　5 pints = _____ cups

16 tablespoons = _____ cup　　　　　　2 pints = _____ quart

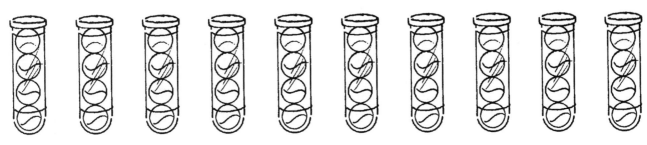

1.

1	2	3	4	5	6	7	8	9
× 4	× 4	× 4	× 4	× 4	× 4	× 4	× 4	× 4

2. 3 × 2 = ____   0 × 3 = ____   4 × 5 = ____   5 × 4 = ____

3. 4 × 4 = ____   3 × 7 = ____   6 × 3 = ____   3 × 3 = ____

4. 3× 4 = ____   7 × 4 = ____   5 × 3 = ____   6 × 4 = ____

5. 8 × 3 = ____   2 × 4 = ____   9 × 4 = ____   6 × 5 = ____

6. 5 × 2 = ____   8 × 4 = ____   6 × 2 = ____   7 × 5 = ____

7.

2 1	4 7	4 1	3 3	6 3
× 4	× 4	× 5	× 3	× 4

8.

3 8	6 5	8 9	7 3	2 5
× 4	× 4	× 4	× 4	× 4

9. ____ × 4 = 20   ____ × 4 = 8   ____ × 4 = 12

   20 ÷ 4 = ____   8 ÷ 4 = ____   12 ÷ 4 = ____

calendar

NOVEMBER						
Sun.	Mon.	Tue.	Wed.	Thur.	Fri.	Sat.

How many days are in November ? _____

What is the first day of November this year ? _____

What is the last day of November this year ? _____

On what date does Thanksgiving come this year ? _____

What month comes before November ? _____

1. 69
   × 3

2. 47
   × 3

3. 21
   × 3

4. 85
   × 3

5. 60
   × 3

6. 37
   × 3

Add.

7. 6
    7
    4
  + 7

8. 9
    5
    3
  + 7

9. 6
    6
    8
  + 9

10. 8
     2
     5
   + 5

11. 3
     3
     3
   + 7

12. 2
     4
     6
   + 8

13. 3
     9
     1
   + 4

14. 5
     5
     7
   + 3

15. 30
    79
  + 51

16. 68
    52
  + 47

17. 76
    36
  + 58

18. 95
    90
  + 25

19. 63
    26
  + 43

20. 28
    32
  + 57

Subtract.

21. 23
   − 17

22. 54
   − 29

23. 82
   − 38

24. 33
   − 27

25. 100
   − 65

26. 23
   − 17

42

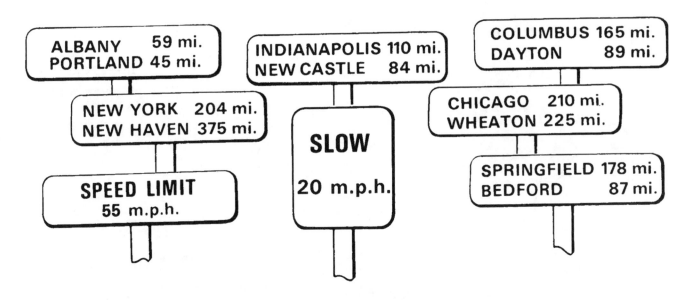

1. How far is it to Albany ? _____

2. How far is it to New Castle ? _____

3. What city is the farthest away ? _____

4. Which city is closer, Chicago or New York ? _____

5. What is the speed limit ? _____

6.
$$\begin{array}{r} 62 \\ \times\ 4 \\ \hline \end{array}$$
$$\begin{array}{r} 37 \\ \times\ 4 \\ \hline \end{array}$$
$$\begin{array}{r} 86 \\ \times\ 4 \\ \hline \end{array}$$
$$\begin{array}{r} 72 \\ \times\ 4 \\ \hline \end{array}$$
$$\begin{array}{r} 95 \\ \times\ 4 \\ \hline \end{array}$$

7.
$$\begin{array}{r} 21 \\ \times 34 \\ \hline \end{array}$$
$$\begin{array}{r} 63 \\ \times 53 \\ \hline \end{array}$$
$$\begin{array}{r} 84 \\ \times 23 \\ \hline \end{array}$$
$$\begin{array}{r} 60 \\ \times 25 \\ \hline \end{array}$$
$$\begin{array}{r} 72 \\ \times 24 \\ \hline \end{array}$$

8. $4\overline{)28}$ $4\overline{)12}$ $4\overline{)20}$ $4\overline{)36}$ $4\overline{)32}$

THANKSGIVING SALES

TURKEY
$1.39 lb.

FROZEN CORN
$1.45

POTATOES
5 lb. bag $1.99

CRANBERRY SAUCE $.79

PUMPKIN PIE $3.50

1. How much would a bag of potatoes and a box of corn cost ? _____

2. How much would 2 pies cost ? _____

3. You have $5.00. You buy a pie. What is your change ? _____

4. Four cans of cranberry sauce cost _____ .

Multiply.

5.
$$565 \times 4$$
$$874 \times 4$$
$$425 \times 5$$
$$641 \times 4$$

6.
$$43 \times 35$$
$$78 \times 34$$
$$52 \times 54$$
$$68 \times 45$$

7. $4\overline{)16}$ $4\overline{)8}$ $4\overline{)40}$ $4\overline{)44}$

The Romans used numbers to count too.
Their numerals looked like this:

I	V	X	L	C	D	M
1	5	10	50	100	500	1,000

It was not easy to count or add with Roman numerals.

How do you think the Romans would write 2 ? _____

Try to write these numerals the way the Romans did.

3 _____   4 _____   6 _____   15 _____   20 _____   24 _____

30 _____   36 _____   40 _____   55 _____   110 _____   1,269 _____

Write these Roman numerals as our numerals.

V _____   IX _____   X _____   XIV _____   XVII _____   L _____

XXII _____   XXVI _____   XXIX _____   XXXIII _____   XXXVI _____   M _____

Multiply.

1. $\begin{array}{r} 52 \\ \times\ 5 \\ \hline \end{array}$ $\begin{array}{r} 34 \\ \times\ 5 \\ \hline \end{array}$ $\begin{array}{r} 16 \\ \times\ 5 \\ \hline \end{array}$ $\begin{array}{r} 78 \\ \times\ 5 \\ \hline \end{array}$ $\begin{array}{r} 90 \\ \times\ 5 \\ \hline \end{array}$ $\begin{array}{r} 64 \\ \times\ 5 \\ \hline \end{array}$

2. $\begin{array}{r} 72 \\ \times\ 2 \\ \hline \end{array}$ $\begin{array}{r} 35 \\ \times\ 2 \\ \hline \end{array}$ $\begin{array}{r} 14 \\ \times\ 2 \\ \hline \end{array}$ $\begin{array}{r} 69 \\ \times\ 2 \\ \hline \end{array}$ $\begin{array}{r} 80 \\ \times\ 2 \\ \hline \end{array}$ $\begin{array}{r} 46 \\ \times\ 2 \\ \hline \end{array}$

3. $\begin{array}{r} 79 \\ \times\ 23 \\ \hline \end{array}$ $\begin{array}{r} 82 \\ \times\ 43 \\ \hline \end{array}$ $\begin{array}{r} 37 \\ \times\ 53 \\ \hline \end{array}$ $\begin{array}{r} 53 \\ \times\ 23 \\ \hline \end{array}$ $\begin{array}{r} 16 \\ \times\ 53 \\ \hline \end{array}$ $\begin{array}{r} 40 \\ \times\ 43 \\ \hline \end{array}$

Divide.

4. $4\overline{)28}$ $5\overline{)25}$ $5\overline{)35}$ $8\overline{)24}$ $6\overline{)18}$ $2\overline{)18}$

5. $4\overline{)36}$ $2\overline{)8}$ $5\overline{)40}$ $2\overline{)16}$ $3\overline{)12}$ $3\overline{)9}$

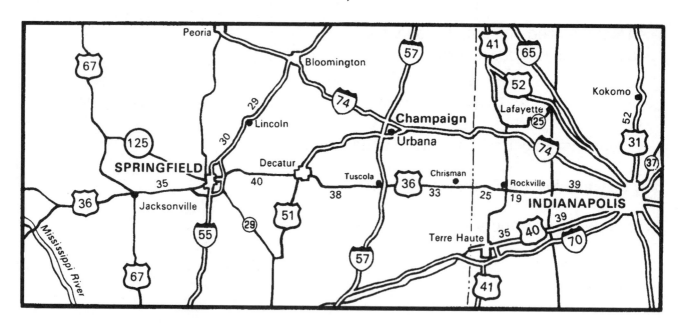

How are highway numbers marked ? _____

Write the name of a city on highway 36. _____

What highways go through Jacksonville ? _____

Name a highway that goes through Lafayette. _____

What two cities seem to have the most highways ? _____ _____

How far is it from Springfield to Decatur ? _____

How far is it from Decatur to Tuscola ? _____

How far is it from Tuscola to Rockville ? _____

How far is it from Rockville to Indianapolis ? _____

How far is it from Springfield to Indianapolis ? _____

How far is it from Indianapolis to Terre Haute ? _____

How far is it from Springfield to Jacksonville ? _____

How far is it from Springfield to Bloomington ? _____

Subtract.

1. $\begin{array}{r} 76 \\ -37 \\ \hline \end{array}$	2. $\begin{array}{r} 80 \\ -25 \\ \hline \end{array}$	3. $\begin{array}{r} 58 \\ -23 \\ \hline \end{array}$	4. $\begin{array}{r} 41 \\ -19 \\ \hline \end{array}$	5. $\begin{array}{r} 22 \\ -22 \\ \hline \end{array}$
6. $\begin{array}{r} 37 \\ -29 \\ \hline \end{array}$	7. $\begin{array}{r} 60 \\ -50 \\ \hline \end{array}$	8. $\begin{array}{r} 93 \\ -54 \\ \hline \end{array}$	9. $\begin{array}{r} 47 \\ -30 \\ \hline \end{array}$	10. $\begin{array}{r} 65 \\ -37 \\ \hline \end{array}$
11. $\begin{array}{r} 59 \\ -29 \\ \hline \end{array}$	12. $\begin{array}{r} 40 \\ -36 \\ \hline \end{array}$	13. $\begin{array}{r} 87 \\ -39 \\ \hline \end{array}$	14. $\begin{array}{r} 50 \\ -40 \\ \hline \end{array}$	15. $\begin{array}{r} 60 \\ -17 \\ \hline \end{array}$
16. $\begin{array}{r} 45 \\ -25 \\ \hline \end{array}$	17. $\begin{array}{r} 95 \\ -56 \\ \hline \end{array}$	18. $\begin{array}{r} 90 \\ -30 \\ \hline \end{array}$	19. $\begin{array}{r} 82 \\ -39 \\ \hline \end{array}$	20. $\begin{array}{r} 47 \\ -45 \\ \hline \end{array}$
21. $\begin{array}{r} 92 \\ -35 \\ \hline \end{array}$	22. $\begin{array}{r} 34 \\ -19 \\ \hline \end{array}$	23. $\begin{array}{r} 70 \\ -20 \\ \hline \end{array}$	24. $\begin{array}{r} 76 \\ -68 \\ \hline \end{array}$	25. $\begin{array}{r} 58 \\ -7 \\ \hline \end{array}$
26. $\begin{array}{r} 66 \\ -47 \\ \hline \end{array}$	27. $\begin{array}{r} 80 \\ -36 \\ \hline \end{array}$	28. $\begin{array}{r} 73 \\ -43 \\ \hline \end{array}$	29. $\begin{array}{r} 59 \\ -34 \\ \hline \end{array}$	30. $\begin{array}{r} 93 \\ -47 \\ \hline \end{array}$
31. $\begin{array}{r} 52 \\ -21 \\ \hline \end{array}$	32. $\begin{array}{r} 61 \\ -41 \\ \hline \end{array}$	33. $\begin{array}{r} 57 \\ -54 \\ \hline \end{array}$	34. $\begin{array}{r} 97 \\ -29 \\ \hline \end{array}$	35. $\begin{array}{r} 20 \\ -10 \\ \hline \end{array}$
36. $\begin{array}{r} 75 \\ -30 \\ \hline \end{array}$	37. $\begin{array}{r} 83 \\ -75 \\ \hline \end{array}$	38. $\begin{array}{r} 75 \\ -8 \\ \hline \end{array}$	39. $\begin{array}{r} 90 \\ -47 \\ \hline \end{array}$	40. $\begin{array}{r} 64 \\ -64 \\ \hline \end{array}$

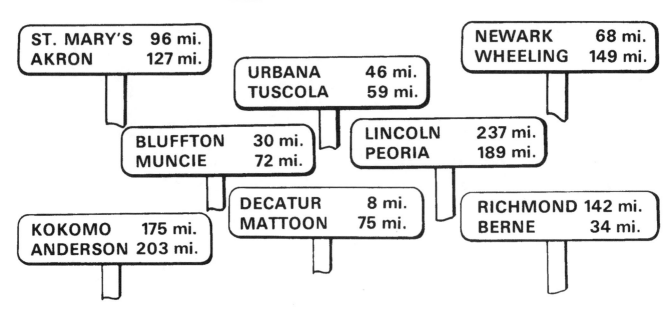

ST. MARY'S 96 mi.
AKRON   127 mi.

URBANA   46 mi.
TUSCOLA 59 mi.

NEWARK   68 mi.
WHEELING 149 mi.

BLUFFTON 30 mi.
MUNCIE   72 mi.

LINCOLN 237 mi.
PEORIA   189 mi.

KOKOMO   175 mi.
ANDERSON 203 mi.

DECATUR 8 mi.
MATTOON 75 mi.

RICHMOND 142 mi.
BERNE   34 mi.

1. How far is it to Peoria ? _____

2. How far is it to Akron ? _____

3. Which is farther, Peoria or Akron ? _____

4. Which city is the closest of all ? _____

5. How far is it to Kokomo ? _____

6.
$$42 + 87$$ $$61 + 49$$ $$23 + 18$$ $$76 + 15$$ $$84 + 48$$

7.
$$82 - 45$$ $$63 - 28$$ $$22 - 14$$ $$30 - 12$$ $$432 - 141$$

8.
$$41 \times 3$$ $$52 \times 4$$ $$70 \times 4$$ $$83 \times 5$$ $$47 \times 2$$

9.
$$4)\overline{8}$$ $$4)\overline{20}$$ $$5)\overline{35}$$ $$3)\overline{15}$$ $$2)\overline{18}$$

Crazy Chocolate Cake

1 1/2 cups sifted flour
3 tablespoons baking cocoa
1 teaspoon baking soda
1 cup sugar
1/2 teaspoon salt

5 tablespoons oil
1 tablespoon vinegar
1 teaspoon vanilla
1 cup cold water

Sift flour, cocoa, baking soda, sugar, and salt together into 9" x 9" square pan. Make three holes in the mixture. Put oil in one hole, vinegar in another and vanilla in the third hole. Pour cold water over the whole thing and mix until flour disappears. Bake at 350° for 30 minutes.
Frost if desired.

Multiply.

1.	46	90	87	32	14
	× 2	× 3	× 5	× 2	× 3

2.	57	10	35	89	35
	× 3	× 3	× 5	× 2	× 3

3.	53	98	53	19	50
	× 25	× 13	× 22	× 25	× 32

Divide.

4. $5\overline{)15}$ $4\overline{)12}$ $3\overline{)12}$ $2\overline{)12}$ $3\overline{)18}$

5. $3\overline{)24}$ $3\overline{)27}$ $4\overline{)36}$ $2\overline{)18}$ $3\overline{)18}$

Molded Strawberry Salad

2 packages strawberry gelatin (3 oz each)
1 can crushed pineapple (20 oz)
1 lb package frozen strawberries
1/2 cup chopped walnuts
pinch of salt

Heat pineapple in a saucepan, juice and all.
Stir in gelatin until it is all dissolved.
Add strawberries, nuts and salt.
Pour into pan or bowl and cool in refrigerator until set.
Top with whipped topping, if desired.

Subtract.

1. 82 − 76	**2.** 79 − 6	**3.** 80 − 43	**4.** 93 − 54	**5.** 39 − 19
6. 74 − 65	**7.** 90 − 20	**8.** 80 − 46	**9.** 63 − 45	**10.** 54 − 40
11. 73 − 73	**12.** 45 − 18	**13.** 26 − 24	**14.** 32 − 15	**15.** 93 − 74
16. 70 − 64	**17.** 78 − 30	**18.** 90 − 87	**19.** 87 − 3	**20.** 75 − 19
21. 30 − 20	**22.** 83 − 36	**23.** 94 − 65	**24.** 70 − 40	**25.** 82 − 58
26. 91 − 89	**27.** 45 − 36	**28.** 60 − 57	**29.** 47 − 43	**30.** 85 − 24

Handmade Cookies

1 cup brown sugar
1 cup flour
1 cup margarine
2 cups oatmeal
1 teaspoon baking soda

Put all ingredients in a large bowl. Squish, mash, pound, and knead by hand until mixed. Form into small balls and place on an ungreased cookie sheet. Bake at 350° for 10-12 minutes. Let cool before removing from cookie sheet.

Makes 4 dozen.

1.

```
   55        99        79        27        98
   46        86        62        46        37
 + 57      + 69      + 75      + 87      + 70
```

2.

```
   42        36        47        83        90
 ×  5      ×  5      ×  4      ×  4      ×  5
```

3.

```
  761       804       305       723       814
 ×   3     ×   4     ×   3     ×   4     ×   3
```

4.

```
   83        61        80        39        47
 × 54      × 42      × 24      × 55      × 51
```

Chapter 3 *Multiplication and Division*

Multiply.

0	1	2	3	4	5	6	7	8	9	10
× 4	× 4	× 4	× 4	× 4	× 4	× 4	× 4	× 4	× 4	× 4

1. $2 \times 4 =$ ____ $3 \times 4 =$ ____ $4 \times 1 =$ ____ $4 \times 0 =$ ____ $4 \times 5 =$ ____

2. $4 \times 2 =$ ____ $4 \times 3 =$ ____ $1 \times 4 =$ ____ $0 \times 4 =$ ____ $5 \times 4 =$ ____

3.

21	36	48	50	79	92
× 4	× 4	× 4	× 4	× 4	× 4

4.

43	56	28	60	85	91
× 5	× 5	× 5	× 5	× 5	× 5

5.

24	40	68	35	17	92
× 3	× 3	× 3	× 3	× 3	× 3

6.

19	32	94	56	88	78
× 2	× 2	× 2	× 2	× 2	× 2

7.

58	67	93	70	42	73
× 24	× 34	× 54	× 53	× 34	× 25

8. I have four quarters. How much money do I have? _____

multiplication

Write the multiplication facts for 4.

Multiply.

1. $\begin{array}{r} 25 \\ \times\ 4 \\ \hline \end{array}$ $\begin{array}{r} 63 \\ \times\ 5 \\ \hline \end{array}$ $\begin{array}{r} 87 \\ \times\ 2 \\ \hline \end{array}$ $\begin{array}{r} 54 \\ \times\ 3 \\ \hline \end{array}$ $\begin{array}{r} 90 \\ \times\ 4 \\ \hline \end{array}$ $\begin{array}{r} 15 \\ \times\ 5 \\ \hline \end{array}$

2. $\begin{array}{r} 20 \\ \times\ 2 \\ \hline \end{array}$ $\begin{array}{r} 37 \\ \times\ 3 \\ \hline \end{array}$ $\begin{array}{r} 49 \\ \times\ 4 \\ \hline \end{array}$ $\begin{array}{r} 76 \\ \times\ 4 \\ \hline \end{array}$ $\begin{array}{r} 83 \\ \times\ 5 \\ \hline \end{array}$ $\begin{array}{r} 67 \\ \times\ 3 \\ \hline \end{array}$

3. $\begin{array}{r} 24 \\ \times\ 2 \\ \hline \end{array}$ $\begin{array}{r} 50 \\ \times\ 4 \\ \hline \end{array}$ $\begin{array}{r} 19 \\ \times\ 2 \\ \hline \end{array}$ $\begin{array}{r} 31 \\ \times\ 5 \\ \hline \end{array}$ $\begin{array}{r} 28 \\ \times\ 2 \\ \hline \end{array}$ $\begin{array}{r} 46 \\ \times\ 4 \\ \hline \end{array}$

4. $\begin{array}{r} 73 \\ \times\ 3 \\ \hline \end{array}$ $\begin{array}{r} 92 \\ \times\ 5 \\ \hline \end{array}$ $\begin{array}{r} 80 \\ \times\ 4 \\ \hline \end{array}$ $\begin{array}{r} 56 \\ \times\ 3 \\ \hline \end{array}$ $\begin{array}{r} 39 \\ \times\ 2 \\ \hline \end{array}$ $\begin{array}{r} 17 \\ \times\ 4 \\ \hline \end{array}$

5. $\begin{array}{r} 10 \\ \times\ 5 \\ \hline \end{array}$ $\begin{array}{r} 68 \\ \times\ 4 \\ \hline \end{array}$ $\begin{array}{r} 53 \\ \times\ 5 \\ \hline \end{array}$ $\begin{array}{r} 85 \\ \times\ 3 \\ \hline \end{array}$ $\begin{array}{r} 27 \\ \times\ 2 \\ \hline \end{array}$ $\begin{array}{r} 98 \\ \times\ 4 \\ \hline \end{array}$

6. $\begin{array}{r} 57 \\ \times\ 3 \\ \hline \end{array}$ $\begin{array}{r} 40 \\ \times\ 5 \\ \hline \end{array}$ $\begin{array}{r} 36 \\ \times\ 4 \\ \hline \end{array}$ $\begin{array}{r} 18 \\ \times\ 5 \\ \hline \end{array}$ $\begin{array}{r} 75 \\ \times\ 3 \\ \hline \end{array}$ $\begin{array}{r} 59 \\ \times\ 4 \\ \hline \end{array}$

7. $\begin{array}{r} 82 \\ \times 23 \\ \hline \end{array}$ $\begin{array}{r} 91 \\ \times 35 \\ \hline \end{array}$ $\begin{array}{r} 64 \\ \times 44 \\ \hline \end{array}$ $\begin{array}{r} 47 \\ \times 12 \\ \hline \end{array}$ $\begin{array}{r} 30 \\ \times 54 \\ \hline \end{array}$ $\begin{array}{r} 29 \\ \times 35 \\ \hline \end{array}$

×	1	2	3	4	5
1					
2					
3					
4					
5					
6					
7					
8					
9					
0					

Multiply.

1. 25
    × 5

2. 37
    × 2

3. 15
    × 4

4. 48
    × 3

5. 56
    × 10

6. 79
    × 31

Divide.

$\dfrac{4}{4\overline{)16}}$ How many 4s in 16 ?

$4\overline{)12}$ How many 4s does it take to make 12 ? _____

$4\overline{)20}$ 4 × _____ = 20   How many 4s in 20 ? _____

7.   5 × 5 = _____ $5\overline{)25}$ 5 × 8 = _____ $5\overline{)40}$

8.   3 × 9 = _____ $3\overline{)27}$ 3 × 6 = _____ $3\overline{)18}$

9.   2 × 7 = _____ $2\overline{)14}$ 2 × 2 = _____ $2\overline{)4}$

10. 4 × 8 = _____ $4\overline{)32}$ 4 × 6 = _____ $4\overline{)24}$

11. 5 × 4 = _____ $5\overline{)20}$ 4 × 7 = _____ $4\overline{)28}$

1. How would you count three dimes ? _____ _____ _____

2. Another way to count three dimes is to say 3 × 10¢ = _____.

3. Now count five nickels. _____ _____ _____ _____ _____

4. Then five nickels equals 5 × 5¢ = _____ .

Multiply.

5.	10¢ × 6

6.	12¢ × 5

7.	16¢ × 3

```
  27¢
× 12
  54
27
324¢

324¢ =$3.24
```

8.　13¢　　　　**9.**　18¢　　　　**10.**　20¢
　　　× 4　　　　　　　× 2　　　　　　× 32

11.　15¢　　　　**12.**　17¢　　　　**13.**　36¢
　　　× 45　　　　　　× 13　　　　　　× 24

14.　19¢　　**15.**　34¢　　**16.**　42¢　　**17.**　87¢　　**18.**　63¢
　　× 5　　　　　× 4　　　　× 20　　　　× 5　　　　× 5

Divide and check.

19. 5)‾2‾0‾　　5 × _____ = 20　　**20.** 3)‾1‾8‾

21. 5)‾4‾5‾　　　　　　　　　　　　**22.** 4)‾3‾6‾

23. 3)‾2‾1‾　　　　　　　　　　　　**24.** 3)‾2‾4‾

25. 5)‾3‾0‾　　　　　　　　　　　　**26.** 2)‾1‾0‾

Baked Apples

Use an electric frying pan.
Wash 6 apples, core, pare 1/3 way down from the stem end.
Place in an 8" x 8" x 2" pan. Combine 1/2 cup granulated
sugar, 1/4 cup brown sugar, 1/2 cup water. Pour over apples.
Add 1 tbs butter. Sprinkle apples with cinnamon
or nutmeg. Set on rack. Cover. Close vent. Set dial
at 420°. Bake about 45 min or until apples are tender
Serve with plain or whipped cream.

1.

48	72	34	28	13	42
× 6	× 7	× 5	× 7	× 9	× 8

2.

242	637	389	451	269
× 7	× 5	× 6	× 4	× 5

3.

52	68	87	98	47
× 76	× 54	× 34	× 50	× 43

Divide and check.

4. $4\overline{)32}$ $3\overline{)27}$ $2\overline{)12}$ $3\overline{)18}$

5. $5\overline{)40}$ $4\overline{)36}$ $3\overline{)15}$ $4\overline{)20}$

Thursday's TV Programs			
P.M.	**WGN** Channel 12	**WOW** Channel 5	**WANB** Channel 8	**WKGJ** Channel 10

5	:00	Local News	Bowling	Bowling	News
	:15		News	Showtime	
	:30	Sports	Weather	Travel Update	Weather
	:45			Showtime	

6	:00	World News	News Wrap-up	News	News, Sports
	:15				News-go-round
	:30	Weather Today	Don Smith Show	Evening Report	Joe Swinger
	:45			Sports Report	Five Star Extra

7	:00	News	News	News	Local News
	:15	Music			
	:30	The Wheel Deal	Sports		World News
	:45				

At what time is Weather Today on WGN ? _____

At what time is Evening Report on WANB ? _____

At what time is Five Star Extra on WKGJ ? _____

At what time is Sports on WOW ? _____

How long does World News last on WGN ? _____

How long does the sports program last on WGN ? _____

At what time of the day would you hear the programs above ?

Morning _____   Noon _____   Afternoon _____   Evening _____

At what time would you look for
the weather report on WOW ? _____

At what time would you look for
the Sports Report on WANB ? _____

One	6	is	_____	$6 \times 1 =$ _____	One	six	is	_____
Two	6s	are	_____	$6 \times 2 =$ _____	Two	sixes	are	_____
Three	6s	are	_____	$6 \times 3 =$ _____	Three	sixes	are	_____
Four	6s	are	_____	$6 \times 4 =$ _____	Four	sixes	are	_____
Five	6s	are	_____	$6 \times 5 =$ _____	Five	sixes	are	_____
Six	6s	are	_____	$6 \times 6 =$ _____	Six	sixes	are	_____
Seven	6s	are	_____	$6 \times 7 =$ _____	Seven	sixes	are	_____
Eight	6s	are	_____	$6 \times 8 =$ _____	Eight	sixes	are	_____
Nine	6s	are	_____	$6 \times 9 =$ _____	Nine	sixes	are	_____

1. $\begin{array}{r} 1 \\ \times\,6 \\ \hline \end{array}$	2. $\begin{array}{r} 2 \\ \times\,6 \\ \hline \end{array}$	3. $\begin{array}{r} 3 \\ \times\,6 \\ \hline \end{array}$	4. $\begin{array}{r} 4 \\ \times\,6 \\ \hline \end{array}$	5. $\begin{array}{r} 5 \\ \times\,6 \\ \hline \end{array}$	6. $\begin{array}{r} 6 \\ \times\,6 \\ \hline \end{array}$
7. $\begin{array}{r} 29 \\ \times\,6 \\ \hline \end{array}$	8. $\begin{array}{r} 34 \\ \times\,6 \\ \hline \end{array}$	9. $\begin{array}{r} 15 \\ \times\,6 \\ \hline \end{array}$	10. $\begin{array}{r} 41 \\ \times\,6 \\ \hline \end{array}$	11. $\begin{array}{r} 20 \\ \times\,6 \\ \hline \end{array}$	12. $\begin{array}{r} 85 \\ \times\,6 \\ \hline \end{array}$
13. $\begin{array}{r} 73 \\ \times\,6 \\ \hline \end{array}$	14. $\begin{array}{r} 64 \\ \times\,6 \\ \hline \end{array}$	15. $\begin{array}{r} 52 \\ \times\,6 \\ \hline \end{array}$	16. $\begin{array}{r} 87 \\ \times\,6 \\ \hline \end{array}$	17. $\begin{array}{r} 19 \\ \times\,6 \\ \hline \end{array}$	18. $\begin{array}{r} 63 \\ \times\,6 \\ \hline \end{array}$
19. $\begin{array}{r} 35 \\ \times\,6 \\ \hline \end{array}$	20. $\begin{array}{r} 54 \\ \times\,6 \\ \hline \end{array}$	21. $\begin{array}{r} 21 \\ \times\,6 \\ \hline \end{array}$	22. $\begin{array}{r} 70 \\ \times\,6 \\ \hline \end{array}$	23. $\begin{array}{r} 46 \\ \times\,6 \\ \hline \end{array}$	24. $\begin{array}{r} 92 \\ \times\,6 \\ \hline \end{array}$
25. $\begin{array}{r} 55 \\ \times\,6 \\ \hline \end{array}$	26. $\begin{array}{r} 78 \\ \times\,6 \\ \hline \end{array}$	27. $\begin{array}{r} 37 \\ \times\,6 \\ \hline \end{array}$	28. $\begin{array}{r} 49 \\ \times\,6 \\ \hline \end{array}$	29. $\begin{array}{r} 24 \\ \times\,6 \\ \hline \end{array}$	30. $\begin{array}{r} 95 \\ \times\,6 \\ \hline \end{array}$
31. $\begin{array}{r} 16 \\ \times\,6 \\ \hline \end{array}$	32. $\begin{array}{r} 60 \\ \times\,6 \\ \hline \end{array}$	33. $\begin{array}{r} 83 \\ \times\,6 \\ \hline \end{array}$	34. $\begin{array}{r} 18 \\ \times\,6 \\ \hline \end{array}$	35. $\begin{array}{r} 47 \\ \times\,6 \\ \hline \end{array}$	36. $\begin{array}{r} 86 \\ \times\,6 \\ \hline \end{array}$

This is Jim.

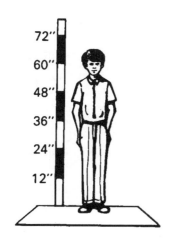

12 inches	= 1 foot
12 in	= 1 ft
twelve inches	= one foot
12"	= 1'
3 feet	= 1 yard
three feet	= one yard
3'	= 1 yd
3 ft	= 1 yd

About how many inches tall is Jim ? _____

About how many feet tall is Jim ? _____

How many inches tall are you ? _____

How many feet tall are you ? _____

How many inches long is this ruler ? _____

Show the inches on this ruler.

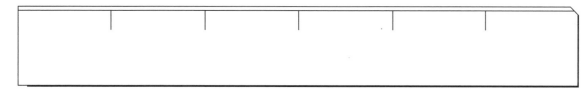

Draw a ruler 6 inches long. Mark the inches.

Draw another ruler 6 inches long. Mark the inches and $\frac{1}{2}$ inches.

Draw a line segment that measures:

$4\frac{1}{2}$ inches

Multiply.

1.	2 2	2.	3 5	3.	7 1	4.	9 4	5.	5 0	6.	4 2
	× 6		× 6		× 6		× 6		× 6		× 6

7.	6 4	8.	8 7	9.	9 3	10.	4 4	11.	5 8	12.	3 0
	× 6		× 6		× 6		× 6		× 6		× 6

13.	8 5	14.	6 9	15.	1 4	16.	9 7	17.	2 8	18.	4 5
	× 6		× 6		× 6		× 6		× 6		× 6

19.	7 6	20.	5 3	21.	5 2	22.	3 5	23.	8 7	24.	6 9
	× 6		× 6		× 2 6		× 6 3		× 1 6		× 6 4

25.	1 5	26.	9 4	27.	2 6	28.	4 8	29.	7 9	30.	5 4
	× 4 6		× 6 4		× 6 8		× 7 6		× 3 6		× 9 6

Subtract.

31.	2 5	32.	3 5	33.	8 1	34.	9 4	35.	5 0	36.	7 2
	− 1 5		− 3 0		− 6 5		− 6 9		− 1 4		− 4 7

Divide and check.

37. $6\overline{)48}$ 38. $6\overline{)30}$ 39. $6\overline{)18}$

40. $6\overline{)54}$ 41. $6\overline{)36}$ 42. $6\overline{)6}$

Measure these line segments. Write the measurement on the line segment.

1.

2.

3.

4.

Draw line segments that measure:

4 inches

$3\frac{1}{2}$ inches

Make a calendar for December.

DECEMBER						
Sun.	Mon.	Tue.	Wed.	Thur.	Fri.	Sat.

How many days are there in December ? _____

What day is the first day of December this year ? _____

What day is the last day of December this year ? _____

On what day does Christmas come this year ? _____

What year will it be on the first day of January ? _____

Draw line segments that measure:

6 inches

$1\frac{1}{2}$ inches

1. 42 77 + 51	**2.** 22 13 + 65	**3.** 98 10 + 67	**4.** 34 46 + 81	**5.** 50 27 + 92	**6.** 24 42 + 79
7. 36 44 + 75	**8.** 29 61 + 15	**9.** 42 57 + 31	**10.** 44 18 + 13	**11.** 83 23 + 25	**12.** 12 1 + 45

Multiply.

```
    42
  × 51
    42
  210
  2142
```

13. 60 × 10

14. 53 × 40

15. 72 × 36

16. 25 × 64

Divide and check.

17. 9)54

19. 3)27

19. 7)28

20. 9)45

21. 5)5

22. 7)35

23. 3)18

24. 8)48

25. 4)12

How much is two 7s ? _____   How much is six 7s ? _____

How much is three 7s ? _____   How much is seven 7s ? _____

How much is four 7s ? _____   How much is eight 7s ? _____

How much is five 7s ? _____   How much is nine 7s ? _____

Write the multiplication
facts for 7.

Divide and check.

1. $7\overline{)63}$ $6\overline{)42}$ $7\overline{)28}$ $2\overline{)48}$

2. $8\overline{)56}$ $7\overline{)77}$ $3\overline{)21}$ $2\overline{)12}$

3. $7\overline{)35}$ $7\overline{)14}$ $4\overline{)48}$ $4\overline{)24}$

4. $2\overline{)26}$ $7\overline{)7}$ $6\overline{)48}$ $4\overline{)32}$

5. $5\overline{)45}$ $4\overline{)36}$ $3\overline{)27}$ $5\overline{)40}$

6. $3\overline{)15}$ $5\overline{)35}$ $4\overline{)44}$ $6\overline{)36}$

7. $6\overline{)66}$ $7\overline{)49}$ $3\overline{)24}$ $6\overline{)48}$

8. $6\overline{)54}$ $3\overline{)18}$ $2\overline{)18}$ $4\overline{)28}$

Measure these line segments. Write the measurement on the line segment.

1. └──┘

2. └──┘

3. └────────────────────┘

4. └──────────────────────────┘

5. └──────────────────────────────────────┘

6. └──┘

7. └──────────────┘

8. └────────────────────┘

9. └──┘

10. └──┘

Draw line segments that measure:

3 inches

$3\frac{1}{2}$ inches

7 inches

$6\frac{1}{2}$ inches

3	2	8	6	1	4	7	5	9	10
×7	×7	×7	×7	×7	×7	×7	×7	×7	×7

1. 7)21   7)56   7)42   7)35   7)63

2. 7)70   7)7   7)14   7)28   7)49

3.
```
   43         62         81         73         42
 × 7        × 7        × 6        × 6        × 7
```

4.
```
   34         62         86         49         57
 ×75        ×37        ×72        ×76        ×79
```

16 ounces = 1 pound	2000 pounds = 1 ton

Name one thing you might buy by the ton. _____

How much do you weigh ? _____

How much does this book weigh ? _____

Name one thing you might buy by the ounce. _____

Count by 5s to 100.

5																			100

Tell the time.

 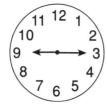   

_____   _____   _____   _____   _____

Write these Roman numerals as our numerals.

IV = _____   IX = _____   XIX = _____   XVII = _____   LV = _____

Multiply.

1.   27   37   44   79   42
         × 3   × 7   × 7   × 6   × 7

Divide.

2. 7⟌43   5⟌37   7⟌16   7⟌39

3. 7⟌30   7⟌25   7⟌52   7⟌60

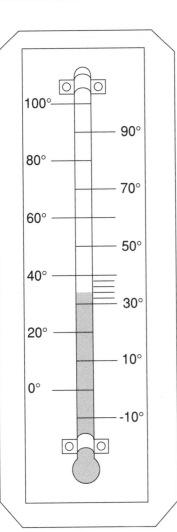

What is the temperature on this thermometer ? _____

What is the temperature on the room thermostat ? _____

What is the temperature on the outdoor thermometer ? _____

What is the normal temperature of your body ? _____

Add.

1. 37	100	97	46
65	75	98	0
+ 49	+ 86	+ 87	+ 32

2. 58	47	108	23
64	39	96	18
+ 69	+ 48	+ 98	+ 15

3. 42	72	49	42
50	26	57	76
+ 24	+ 38	+ 61	+ 64

Divide.

4. $7\overline{)35}$ $6\overline{)48}$ $7\overline{)42}$ $7\overline{)84}$ $7\overline{)63}$

Multiply.

5. 42	72	20	47	61
× 5	× 4	× 6	× 7	× 3

73	**6.** 89	83	32	69
× 21	× 72	× 56	× 74	× 45
73				
146				
1533				

Measure these line segments. Write the measurements on the line segment.

1. └───┘

2. └──┘

3. └──────────────────────┘

4. └────────────────────────────┘

5. └──────────────────────────────────────┘

6. └──┘

7. └──────────────┘

Draw these line segments:

2 inches

$6\frac{1}{2}$ inches

4 inches

3 inches

$5\frac{1}{2}$ inches

1 inch

$1\frac{1}{2}$ inches

Pretzels

Thaw one loaf of frozen bread dough. Let the dough rise according
to package directions.
Divide the dough in half and place on a floured board.
Cut each half into 8 equal pieces.
Roll small pieces into strips that measure 12 inches long and about
1/2 inch around (Adjust according to thickness desired)
Shape into pretzels. Seal with fingers where dough crosses.
Place on a greased baking sheet.
Mix 1 tablespoon of water with a beaten egg yolk. Brush onto top
of pretzels.
Sprinkle with coarse salt.
Bake at 350° for 20-30 minutes until
lightly browned.

Add.

1.	48	604	832	523	393
	+ 27	+ 219	+ 209	+ 764	+ 677

Subtract.

2.	40	82	95	61	47
	− 15	− 14	− 69	− 18	− 38

Multiply.

3.	64	253	34	40	362
	× 7	× 7	× 56	× 37	× 34

Divide.

4. 5)25   2)19   7)50   8)19   5)37

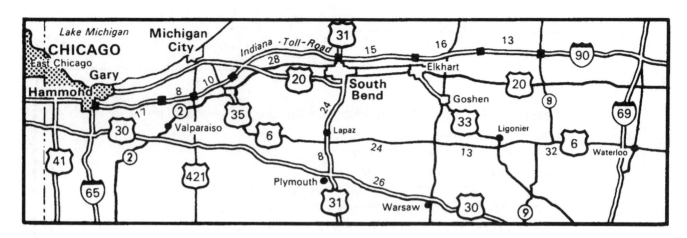

Name two highways that go through South Bend. _____ _____

How far is it from Ligonier to Lapez ? _____

How far is it from Ligonier to Waterloo ? _____

Name a highway that goes through Warsaw. _____

What do the black squares on the Indiana Toll Road mean ?

_____

Name three cities on Lake Michigan. _____ _____

_____

Complete this multiplication chart.

X	0	1	2	3	4	5	6	7	8	9
1										
2										
3										
4										
5										
6										
7										
0										

Use a tape measure.

Measure the number of inches around your waist. _____

Measure the number of inches around your wrist. _____

Measure the length of your sleeve. _____

Measure the height of the windows from the floor. _____

Subtract and check.

| $\begin{array}{r} 42 \\ -24 \\ \hline 18 \end{array} \quad \begin{array}{r} 24 \\ +18 \\ \hline 42 \end{array}$ |

1.
$\begin{array}{r} 87 \\ -49 \\ \hline \end{array}$
$\begin{array}{r} 54 \\ -26 \\ \hline \end{array}$
$\begin{array}{r} 63 \\ -19 \\ \hline \end{array}$

2.
$\begin{array}{r} 41 \\ -37 \\ \hline \end{array}$
$\begin{array}{r} 70 \\ -52 \\ \hline \end{array}$
$\begin{array}{r} 21 \\ -15 \\ \hline \end{array}$
$\begin{array}{r} 96 \\ -40 \\ \hline \end{array}$

3.
$\begin{array}{r} 12 \\ -9 \\ \hline \end{array}$
$\begin{array}{r} 32 \\ -12 \\ \hline \end{array}$
$\begin{array}{r} 91 \\ -45 \\ \hline \end{array}$
$\begin{array}{r} 26 \\ -17 \\ \hline \end{array}$

4.
$\begin{array}{r} 85 \\ -76 \\ \hline \end{array}$
$\begin{array}{r} 50 \\ -30 \\ \hline \end{array}$
$\begin{array}{r} 61 \\ -59 \\ \hline \end{array}$
$\begin{array}{r} 105 \\ -46 \\ \hline \end{array}$

Multiply.

5.
$\begin{array}{r} 53 \\ \times 47 \\ \hline \end{array}$
$\begin{array}{r} 81 \\ \times 36 \\ \hline \end{array}$
$\begin{array}{r} 24 \\ \times 75 \\ \hline \end{array}$
$\begin{array}{r} 60 \\ \times 45 \\ \hline \end{array}$
$\begin{array}{r} 91 \\ \times 20 \\ \hline \end{array}$
$\begin{array}{r} 37 \\ \times 53 \\ \hline \end{array}$

Divide.

6. $7\overline{)48}$ $\qquad$ $5\overline{)39}$ $\qquad$ $2\overline{)5}$ $\qquad$ $3\overline{)25}$ $\qquad$ $4\overline{)29}$

SNACKS

POTATO CHIPS 75¢

CHOCOLATE CHIP COOKIE 65¢

TRAIL MIX 75¢

PEANUT BUTTER CRACKERS 65¢

NUTRI BAR 85¢

POPCORN 85¢

MINTS 35¢

CANDY PIECES 55¢

LEMONADE $1.25

FRUIT PUNCH $1.35

ICE TEA $1.50

WATER $1

Cherry Cobbler

1 No. 2 can sweetened cherries
1 tbs butter
1/4 c sugar
1/4 c water
2 tsp cornstarch
1/4 tsp almond or vanilla extract

1/4 tsp red food coloring
   if desired.
1 package ready-to-bake biscuits
   or 10 small baking
   powder biscuits
sugar and cinnamon

Empty cherries into saucepan. Add butter and sugar.
Set dial at 220°. Combine water and cornstarch, then stir into cherries.
Cook and stir until thickened. Stir in extract and coloring. Arrange
biscuits on top of fruit. Cover, leave vent open slightly. Simmer at
simmering point for 20 min or until bisuits are done. Sprinkle with sugar
and cinnamon. Serve hot with cream, ice cream, whipped cream, or
topping. Sliced peaches, berries, pitted plums, etc. may be used in place
of the cherries. Cooking time about 25 min.
Serves 4 to 6.

1. Two nutri bars cost _____ .

2. How much more does

   popcorn cost than a cookie ? _____

3. How much would potato

   chips, water, and mints cost ? _____

4. Two lemonades, crackers

   and a nutri bar would cost _____ .

5. What coins would you

   use to buy ice tea ? _____

Complete the money chart.

+	1¢	5¢	10¢	25¢	50¢	$1.00
1¢						
5¢						
10¢						
25¢						
50¢						
$1.00						
Total						

1.
$1.20
.35
+ .15

$5.65
1.30
+ .25

$6.48
2.06
+ 1.15

$3.20
1.18
+ 4.35

2.
$7.85
4.98
+ 6.75

$ .25
1.00
+ .34

$ .48
.15
+ .10

$2.30
5.00
+ 7.85

Divide and check.

3. 7)770   7)217   7)63

4. 7)42   5)35   4)444

5. 7)56   8)48   9)54

Draw line segments that measure:

3 inches

$3\frac{1}{2}$ inches

$3\frac{1}{4}$ inches

How much is two 8s ? _____

How much is three 8s ? _____

How much is four 8s ? _____

How much is five 8s ? _____

How much is six 8s ? _____

How much is seven 8s ? _____

How much is eight 8s ? _____

How much is nine 8s ? _____

Write the multiplication facts for 8.

Multiply.

1. 64 × 8	28 × 8	47 × 7	90 × 8	87 × 6	43 × 8
2. 71 × 8	38 × 6	65 × 7	85 × 8	59 × 7	97 × 3
3. 40 × 2	75 × 6	79 × 8	47 × 5	59 × 6	82 × 8
4. 29 × 64	43 × 78	79 × 47	47 × 76	76 × 58	28 × 73

Subtract.

5. 85 − 36	70 − 65	41 − 35	94 − 6
6. 20 − 19	84 − 76	47 − 20	94 − 68

60 seconds = 1 minute	24 hours = 1 day	365 days = 1 year
60 minutes = 1 hour	7 days = 1 week	100 years = 1 century
	12 months = 1 year	

What time is it ? _____

What time will it
be in 25 minutes ? _____

Show where the hands
on the first clock will
be in 40 minutes.

1. 1 century = _____ years

2. 1 week = _____ days

3. 1 year = _____ days

4. 1 minute = _____ seconds

5. 1 day = _____ hours

6. 1 year = _____ months

7. 1 hour = _____ minutes

Add.

8.
```
   63        45        56        52         50
   26        52        30        77         64
 + 89      + 97      + 86     + 129      + 114
```

9.
```
   74       109        43        34         65
   23        36        42        50         32
 + 97      + 73      + 85      + 84       + 26
```

Multiply.

10.
```
   24        59        63        17         80
 ×  8      ×  8      ×  8      ×  8       ×  8
```

11.
```
   56        79        34        50         82
 × 84      × 36      × 78      × 24       × 60
```

1. Measure these line segments. Write the measurement on the line segments.

2. What time is it?

_____   _____   _____   _____   _____

3.  = _____

4. = _____

5.  = _____

6. 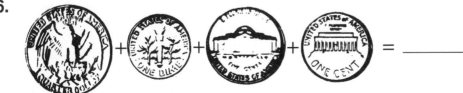 = _____

Subtract.

1.	52 − 4	40 − 3	61 − 7	41 − 3
2.	30 − 7	22 − 9	70 − 8	92 − 8
3.	82 − 5	61 − 2	90 − 6	71 − 5
4.	80 − 9	60 − 2	92 − 6	30 − 1
5.	50 − 4	21 − 6	57 − 5	90 − 8
6.	65 − 64	80 − 24	89 − 63	96 − 42
7.	88 − 83	72 − 59	41 − 38	65 − 65

Multiply.

8.	62 × 85	80 × 73	41 × 65	78 × 48

1. 30 cents = _____ nickels
2. $1.25 = _____ nickels
3. 50 cents = _____ dimes
4. $ .10 = _____ dimes
5. 65 cents = _____ nickels
6. $ .60 = _____ nickels
7. 75 cents = _____ quarters
8. $ .35 = _____ pennies

9.
```
  $.76      $.79      $.47      $.75      $.62
   .85       .84       .98       .59       .96
 + .64     + .68     + .86     + .64     + .89
```

10.
```
  $.64      $.87      $.35      $.74      $.47
   .78       .48       .85       .87       .69
 + .76     + .69     + .73     + .86     + .53
```

11.
```
    76        64        84        98        59
  × 85      × 73      × 47      × 86      × 75
```

12.
```
    96        89        78        40        95
  × 62      × 64      × 46      × 42      × 78
```

13. $3\overline{)96}$ $8\overline{)72}$ $6\overline{)42}$ $5\overline{)40}$ $2\overline{)24}$

14. $8\overline{)56}$ $4\overline{)32}$ $7\overline{)35}$ $8\overline{)64}$ $9\overline{)36}$

Oatmeal Cookies (5 dozen)

3/4 cup shortening, soft	1 teaspoon vanilla
1 cup brown sugar	1 cup sifted flour
1/2 cup granulated sugar	1 teaspoon salt
1 egg	1/2 teaspoon baking soda
1/4 cup water	3 cups oats, uncooked

Place shortening, sugars, egg, water, and vanilla in mixing
bowl; beat thoroughly. Sift together flour, salt, and soda;
add to shortening mixture, mixing well.
Blend in oats, and drop by teaspoonfuls onto greased cookie sheets.
Bake in moderate oven (350°) 12 to 15 minutes.
(For variety, add chopped nuts, raisins, chocolate chips,
or coconut to the the dough.)

Complete this calendar
for January.

JANUARY						
Sun.	Mon.	Tue.	Wed.	Thur.	Fri.	Sat.

On what day is New Year's ? _____

On what day does January end ? _____

How many days are there in January ? _____

On what day is January 25 ? _____

What month comes after January ? _____

Add.
```
    9        9        9        9        9        9        9        9
  + 9        9        9        9        9        9        9        9
           + 9        9        9        9        9        9        9
                    + 9        9        9        9        9        9
                             + 9        9        9        9        9
                                      + 9        9        9        9
                                               + 9        9        9
                                                        + 9        9
                                                                 + 9
```

Two nines   = _____   Seven nines = _____   9 × 0 = _____   9 × 5 = _____

Three nines = _____   Eight nines = _____   9 × 1 = _____   9 × 6 = _____

Four nines = _____   Nine nines   = _____   9 × 2 = _____   9 × 7 = _____

Five nines = _____   Zero nines   = _____   9 × 3 = _____   9 × 8 = _____

Six nines   = _____   9 × 4 = _____   9 × 9 = _____

Multiply.

1.
```
    25          37          14          68          19
  ×  9        ×  9        ×  9        ×  9        ×  9
```

2.
```
    50          72          43          81          96
  ×  9        ×  9        ×  9        ×  9        ×  9
```

3.
```
    54          60          39          82          77
  ×  9        ×  9        ×  9        ×  9        ×  9
```

4.
```
    45          23          18          98          57
  ×  9        ×  9        ×  9        ×  9        ×  9
```

5.
```
    36          87          49          70          60
  ×  9        ×  9        ×  9        ×  9        ×  9
```

Measure these line segments. Write the measurements on the line segments.

Add.

1.	87	53	64	84	27	56
	+ 29	+ 78	+ 93	+ 36	+ 43	+ 29

2.	$1.43	$2.93	$3.72	$6.27	$6.03	$1.43
	+ 2.76	+ 1.45	+ 1.47	+ 2.73	+ 2.78	+ 9.76

Subtract.

3.	73	82	46	37	46	69
	− 15	− 78	− 28	− 19	− 19	− 49

4.	62¢	37¢	49¢	60¢	52¢	48¢
	− 15¢	− 19¢	− 27¢	− 28¢	− 15¢	− 19¢

Multiply.

5.	42	52	60	55	44
	× 53	× 27	× 38	× 98	× 26

Multiply.

1.
3	6	6	8	7	5	7	2	1
×1	×3	×1	×3	×8	×6	×4	×9	×9

2.
5	5	8	7	8	7	3	3	5
×5	×1	×4	×0	×5	×9	×7	×8	×3

3.
9	2	2	9	2	5	6	5	0
×9	×6	×0	×7	×8	×2	×5	×9	×5

4.
8	1	5	1	3	1	7	1	4
×0	×3	×4	×6	×9	×8	×5	×7	×4

5.
1	5	4	9	9	4	8	7	8
×2	×7	×6	×4	×5	×2	×2	×6	×6

6.
4	2	3	3	7	3	8	3	6
×9	×4	×4	×3	×7	×2	×9	×5	×9

7.
8	3	6	4	2	6	3	4	8
×1	×6	×8	×3	×1	×6	×0	×8	×8

8.
5	7	6	9	8	2	2	2	9
×8	×3	×4	×2	×7	×7	×3	×2	×8

9.
9	7	4	2	9	4	6	6	7
×3	×1	×7	×5	×6	×5	×7	×2	×2

Multiply.

1. 803   426   397   204   416
      ×   7   ×   8   ×   6   ×   8   ×   9

2.   42   83   62   45   86
      × 27   × 78   × 23   × 27   × 15

Divide.

| 5 r 2 |
| 5) 2 7   5 × 5 < 2 7 |
| − 2 5 |
| 2 |

3. 5) 2 3   _____ × 5 < 23

4. 2) 1 9   _____ × 2 < 19   5) 3 8   _____ × 5 < 38

5. 3) 2 0   _____ × 3 < 20   6) 3 9   _____ × 6 < 39

Write the time.

_____   _____   _____   _____   _____

Multiply.

1. 62   26   38   50   38
      × 56   × 25   × 23   × 65   × 25

Divide.

2. 6)‾3‾4‾   _____ × 6 < 34   2)‾1‾3‾   _____ × 2 < 13

3. 7)‾4‾5‾   _____ × 7 < 45   5)‾3‾4‾   _____ × 5 < 34

4. 2)‾1‾9‾   _____ × 2 < 19   3)‾1‾9‾   _____ × 3 < 19

Divide.

1. $2\overline{)6}$ $5\overline{)45}$ $8\overline{)72}$ $5\overline{)5}$ $5\overline{)10}$ $2\overline{)12}$

2. $7\overline{)49}$ $7\overline{)35}$ $2\overline{)10}$ $6\overline{)18}$ $8\overline{)56}$ $3\overline{)9}$

3. $4\overline{)28}$ $9\overline{)27}$ $7\overline{)28}$ $9\overline{)81}$ $7\overline{)63}$ $3\overline{)18}$

4. $9\overline{)36}$ $2\overline{)18}$ $6\overline{)12}$ $8\overline{)64}$ $9\overline{)9}$ $4\overline{)20}$

5. $7\overline{)21}$ $5\overline{)25}$ $3\overline{)15}$ $8\overline{)48}$ $6\overline{)24}$ $3\overline{)3}$

6. $4\overline{)12}$ $4\overline{)16}$ $7\overline{)42}$ $2\overline{)16}$ $5\overline{)15}$ $6\overline{)48}$

7. $8\overline{)32}$ $5\overline{)20}$ $9\overline{)18}$ $8\overline{)24}$ $7\overline{)63}$ $2\overline{)14}$

8. $5\overline{)40}$ $3\overline{)6}$ $5\overline{)35}$ $3\overline{)21}$ $6\overline{)54}$ $1\overline{)3}$

9. $7\overline{)14}$ $3\overline{)27}$ $9\overline{)54}$ $2\overline{)4}$ $3\overline{)12}$ $1\overline{)1}$

10. $9\overline{)63}$ $2\overline{)2}$ $9\overline{)72}$ $6\overline{)30}$ $5\overline{)30}$ $6\overline{)36}$

11. $1\overline{)6}$ $4\overline{)24}$ $3\overline{)24}$ $7\overline{)56}$ $6\overline{)42}$ $8\overline{)8}$

12. $9\overline{)45}$ $4\overline{)36}$ $4\overline{)32}$ $8\overline{)40}$ $8\overline{)16}$ $4\overline{)0}$

1. + + + + = _____ ¢

2. + + = _____ ¢

3.

6	6	6	6	9	6	4	25
× 2	× 7	× 3	× 8	× 6	× 5	× 6	× 6

Divide.

4. 6)̄3̄6̄   6)̄4̄8̄   6)̄5̄4̄   6)̄4̄2̄   6)̄2̄4̄

5. 6)̄3̄9̄   6)̄4̄9̄   6)̄5̄8̄   6)̄4̄5̄   6)̄2̄7̄

6. 2)̄1̄3̄   3)̄2̄0̄   5)̄3̄2̄   7)̄4̄5̄   9)̄5̄5̄

+ + + + + + = _____ ¢

7 × 1 = _____   7 × 5 = _____   7 × 9 = _____

7 × 2 = _____   7 × 6 = _____   7 × 10 = _____

7 × 3 = _____   7 × 7 = _____   7 × 0 = _____

7 × 4 = _____   7 × 8 = _____   10 × 7 = _____

Divide and check.

1. 7) 4 4   7) 1 6   6) 4 3

2. 7) 5 6 7   7) 1 1 2   5) 1 8 5

3. 7) 5 8 8   7) 4 1 4   7) 3 8 6

Multiply.

1.

8	8	8	8	8	8	8	9
× 2	× 5	× 1	× 0	× 6	× 3	× 8	× 8

2.

43	521	643	47	52
× 8	× 8	× 8	× 28	× 83

Divide and check.

3. 8)184   8)368   8)496

4. 8)123   8)205   8)625

Show the time for each zone.

Fill in the blanks. Use counting order.

4			
	2		

Show the time for each time zone.

Pacific Time	Mountain Time	Central Time	Eastern Time

90

1. + + + = _____ ¢

2. + + = _____ ¢

3. $9 \times 7 =$ _____ $9 \times 2 =$ _____ $6 \times 9 =$ _____ $1 \times 9 =$ _____

4. $5 \times 9 =$ _____ $9 \times 8 =$ _____ $3 \times 9 =$ _____ $9 \times 4 =$ _____

5. $8 \times 9 =$ _____ $9 \times 10 =$ _____ $9 \times 9 =$ _____ $9 \times 0 =$ _____

Divide and check.

6. $9\overline{)468}$ $\qquad\qquad$ $3\overline{)276}$ $\qquad\qquad$ $3\overline{)124}$

7. $9\overline{)738}$ $\qquad\qquad$ $9\overline{)513}$ $\qquad\qquad$ $9\overline{)360}$

Show the correct time on each clock.

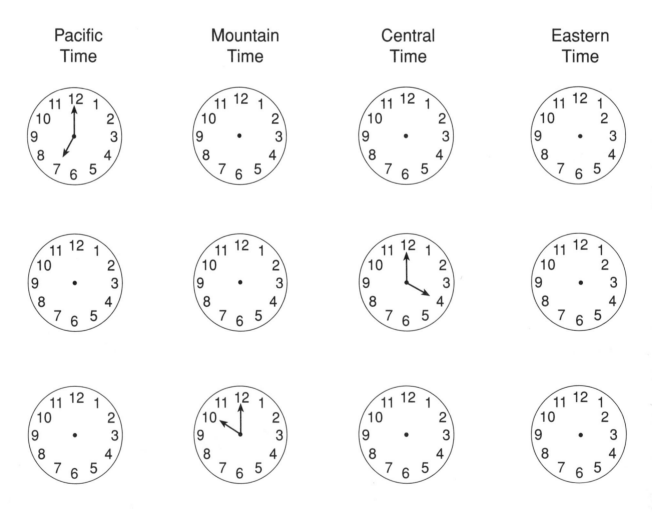

1.  = _____ ¢

2.  = _____ ¢

Multiply.

3.
$$465 \times 3 \qquad 798 \times 2 \qquad 893 \times 4 \qquad 632 \times 5 \qquad 546 \times 3$$

4.
$$45 \times 46 \qquad 89 \times 13 \qquad 26 \times 52 \qquad 28 \times 76 \qquad 47 \times 38$$

Divide and check.

5. $7\overline{)65}$ $\qquad$ $6\overline{)20}$ $\qquad$ $8\overline{)43}$ $\qquad$ $9\overline{)86}$

6. $2\overline{)136}$ $\qquad$ $5\overline{)235}$ $\qquad$ $4\overline{)132}$ $\qquad$ $3\overline{)237}$

You are going to the store. You will pay for each item
with exact change. Complete the chart.

What coins equal these amounts ?	25¢	10¢	5¢	1¢
85¢	3	1		
43¢				
27¢				
51¢				
36¢				

Divide and check.

1. $3\overline{)23}$ $\qquad$ $5\overline{)32}$ $\qquad$ $6\overline{)19}$

2. $2\overline{)56}$ $\qquad$ $3\overline{)78}$ $\qquad$ $5\overline{)67}$

Multiply.

3.
$\begin{array}{r}25\\\times 37\\\hline\end{array}$
$\qquad$
$\begin{array}{r}16\\\times 83\\\hline\end{array}$
$\qquad$
$\begin{array}{r}32\\\times 49\\\hline\end{array}$
$\qquad$
$\begin{array}{r}137\\\times 25\\\hline\end{array}$

You are going to the store. You will pay
for each item with exact change.
Complete the chart.

Items Purchased	Cost	25¢	10¢	5¢	1¢
Nuts	72¢				
Yogurt	93¢				
Juice	68¢				
Milk	52¢				

Multiply.

1. $\begin{array}{r} 24 \\ \times\ 48 \end{array}$ $\begin{array}{r} 64 \\ \times\ 86 \end{array}$ $\begin{array}{r} 20 \\ \times\ 55 \end{array}$ $\begin{array}{r} 54 \\ \times\ 25 \end{array}$ $\begin{array}{r} 98 \\ \times\ 19 \end{array}$

Divide and check.

2. $6\overline{)120}$ $5\overline{)235}$ $6\overline{)327}$

3. $3\overline{)196}$ $2\overline{)435}$ $4\overline{)396}$

4. You buy yogurt. What is
    your change from $1.00 ? _____

1. 10 dimes and 5 pennies = $ _____

2. 8 nickels and 2 pennies = $ _____

3. 2 quarters = $ _____

4. 2 quarters and 2 nickels = $ _____

5. 3 dimes and 4 nickels = $ _____

6. A bagel costs $.55. What
   is your change from $1.00 ? _____

7. How much would 2 bagels cost ? _____

Multiply.

8.
```
   43        84        93        40        83
 × 26      × 27      × 21      × 52      × 26
```

Divide and check.

9. 5)235   2)316   3)174

1. You have $2.25.
   You spend $1.05.
   What is your change ? _____

2. You have $3.32.
   You spend $1.50.
   What is your change ? _____

3. You spent $4.82 at one store.
   You spent $2.48 at the next
   store. How much did you spend ? _____

Multiply.

4.
$$23 \times 46$$
$$87 \times 52$$
$$40 \times 85$$
$$37 \times 93$$

Divide and check.

5. $8\overline{)469}$ $\qquad$ $7\overline{)322}$ $\qquad$ $9\overline{)522}$

Draw line segments that measure:

3 inches

$3\frac{1}{4}$ inches

$2\frac{1}{2}$ inches

$2\frac{1}{4}$ inches

Add.

1.	347 + 26	923 + 147	896 + 203	427 + 847	906 + 138

2.	$3.42 + .63	$1.69 + 2.53	$8.93 + 4.26	$9.38 + 4.27	$3.09 + 4.86

Subtract.

3.	247 − 163	826 − 243	976 − 887	907 − 138	808 − 108

4.	$2.00 − 1.38	$2.38 − 1.19	$6.29 − 1.56	$5.00 − 4.93	$4.27 − 2.83

Fill in each thermometer.

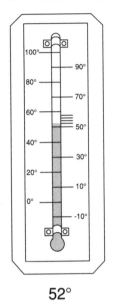 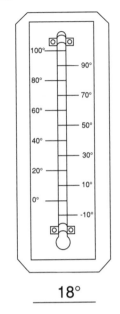 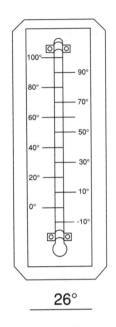 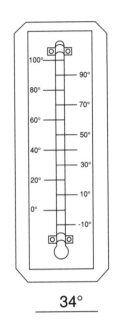

52°   18°   26°   34°

1. You have $1.00.
   You spend $.42.
   What is your change ? _____

2. You have 68¢.
   You spend 39¢.
   What is your change ? _____

3. Pears cost 57¢.
   Peaches cost 63¢. How
   much would both cost ? _____

4. Eggs cost $1.29.
   Bacon costs $3.87.
   What would both cost ? _____

5.
$$\begin{array}{r} 42 \\ \times\ 86 \\ \hline \end{array} \qquad \begin{array}{r} 83 \\ \times\ 27 \\ \hline \end{array} \qquad \begin{array}{r} 48 \\ \times\ 36 \\ \hline \end{array} \qquad \begin{array}{r} 85 \\ \times\ 25 \\ \hline \end{array} \qquad \begin{array}{r} 30 \\ \times\ 56 \\ \hline \end{array}$$

6. $5\overline{)385}$ $7\overline{)315}$ $6\overline{)228}$

What time is it ?

_____   _____   _____   _____   _____

Add.

1.
```
    74          83          46          85          46
  + 29        + 17        + 27        + 98        + 29
```

2.
```
   304         423         923         643         976
 + 278       + 264       + 437       + 293       + 243
```

Subtract.

3.
```
    83          46          87          42          93          64
  - 12        - 24        - 18        - 17        - 45        - 29
```

4.
```
   304         493         641         400         621         846
 - 263       - 287       - 555       - 237       - 419       - 357
```

Find the totals.

1.  = _____ ¢

2.  = _____ ¢

3. Write the temperature.

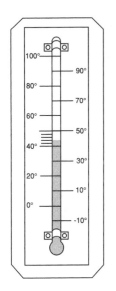 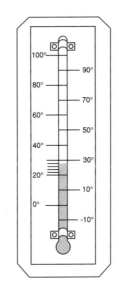 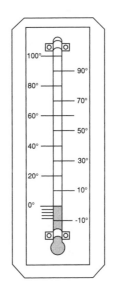 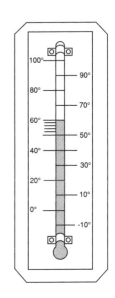

_____   _____   _____   _____

4.
$$\begin{array}{r} 36 \\ \times\, 32 \\ \hline \end{array}$$
$$\begin{array}{r} 47 \\ \times\, 26 \\ \hline \end{array}$$
$$\begin{array}{r} 82 \\ \times\, 53 \\ \hline \end{array}$$
$$\begin{array}{r} 213 \\ \times\quad 5 \\ \hline \end{array}$$
$$\begin{array}{r} 608 \\ \times\quad 7 \\ \hline \end{array}$$

5. 5)235   2)198   3)432   4)248

1. Write the time.

_____   _____   _____   _____   _____

2. Measure these line segments. Write the measurements on the line segments.

Add.

3.
$3.47	$1.29	81	49	347
1.36	.36	2	23	243
+ 2.09	+ 4.38	+ 76	+ 7	+ 127

4.
437	290	106	872	193
+ 268	+ 376	+ 239	+ 309	+ 206

Subtract.

5.
82	63	47	323	287
− 15	− 17	− 19	− 126	− 183

Record these temperatures on a line graph.

Time	Temperature
8:00 A.M.	60°
9:00 A.M.	62°
10:00 A.M.	62°
11:00 A.M.	64°
12:00 Noon	68°
1:00 P.M.	72°
2:00 P.M	76°
3:00 P.M	78°

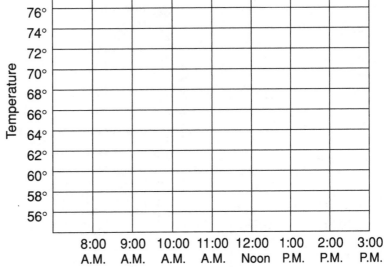

Temperatures During One Day

Multiply.

1.
$$\begin{array}{r} 400 \\ \times\quad 8 \\ \hline \end{array}$$
$$\begin{array}{r} 672 \\ \times\quad 4 \\ \hline \end{array}$$
$$\begin{array}{r} 937 \\ \times\quad 5 \\ \hline \end{array}$$
$$\begin{array}{r} 706 \\ \times\quad 4 \\ \hline \end{array}$$
$$\begin{array}{r} 282 \\ \times\quad 3 \\ \hline \end{array}$$

2.
$$\begin{array}{r} 2812 \\ \times\quad 3 \\ \hline \end{array}$$
$$\begin{array}{r} 6417 \\ \times\quad 5 \\ \hline \end{array}$$
$$\begin{array}{r} 3090 \\ \times\quad 3 \\ \hline \end{array}$$
$$\begin{array}{r} 2172 \\ \times\quad 4 \\ \hline \end{array}$$
$$\begin{array}{r} 3741 \\ \times\quad 2 \\ \hline \end{array}$$

3.
$$\begin{array}{r} 57 \\ \times 49 \\ \hline \end{array}$$
$$\begin{array}{r} 73 \\ \times 24 \\ \hline \end{array}$$
$$\begin{array}{r} 89 \\ \times 37 \\ \hline \end{array}$$
$$\begin{array}{r} 63 \\ \times 54 \\ \hline \end{array}$$
$$\begin{array}{r} 75 \\ \times 56 \\ \hline \end{array}$$

Add.

4.
$$\begin{array}{r} 703 \\ + 549 \\ \hline \end{array}$$
$$\begin{array}{r} 958 \\ + 75 \\ \hline \end{array}$$
$$\begin{array}{r} 780 \\ + 234 \\ \hline \end{array}$$
$$\begin{array}{r} 694 \\ + 546 \\ \hline \end{array}$$
$$\begin{array}{r} 184 \\ + 56 \\ \hline \end{array}$$

Record these temperatures on a line graph.

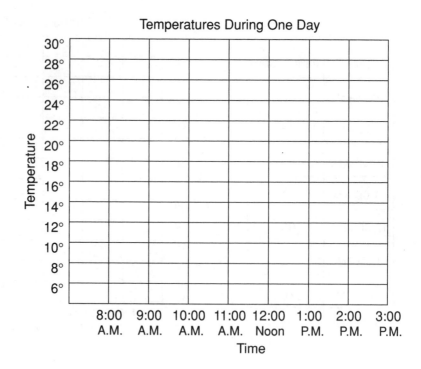

Time	Temperature
8:00 A.M.	28°
9:00 A.M.	24°
10:00 A.M.	23°
11:00 A.M.	22°
12:00 Noon	21°
1:00 P.M.	18°
2:00 P.M.	15°
3:00 P.M.	12°

Multiply.

1.
348	675	932	861	600	382
× 4	× 4	× 3	× 7	× 9	× 8

2.
487	29	39	27	38	49
× 6	× 43	× 54	× 46	× 69	× 73

Add.

3.
144	163	453	255	169	246
372	523	227	128	624	367
+ 409	+ 342	+ 373	+ 464	+ 537	+ 248

Show these temperatures on a line graph.

1 P.M.	70	5 P.M.	71	9 P.M.	60	1 A.M.	58	5 A.M.	56	9 A.M.	60
2 P.M.	72	6 P.M.	68	10 P.M.	59	2 A.M.	57	6 A.M.	57	10 A.M.	64
3 P.M.	73	7 P.M.	65	11 P.M.	59	3 A.M.	56	7 A.M.	58	11 A.M.	69
4 P.M.	73	8 P.M.	62	Midnight	59	4 A.M.	57	8 A.M.	59	Noon	70

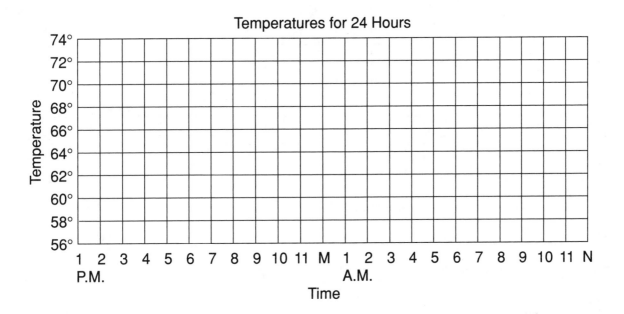

Temperatures for 24 Hours

1. What was the lowest temperature this day ? _____ What was the highest ? _____

Add.

2.
```
    7        28       247      2379       326       458
    6        49       629        56       714       263
    8        63       324      2438       528       462
  + 5      + 77     + 482     + 877       712       172
                                        + 501     + 108
```

Subtract.

3.
```
  409       82        70       397       875
 - 36     - 47      - 34     - 179      - 73
```

line graph

What was the temperature at each hour ?

1 P.M. _____

2 P.M. _____

3 P.M. _____

4 P.M. _____

5 P.M. _____

6 P.M. _____

7 P.M. _____

8 P.M. _____

9 P.M. _____

10 P.M. _____

11 P.M. _____

Midnight _____

Temperatures for 24 Hours

1 A.M. _____

2 A.M. _____

3 A.M. _____

4 A.M. _____

5 A.M. _____

6 A.M. _____

7 A.M. _____

8 A.M. _____

9 A.M. _____

10 A.M. _____

11 A.M. _____

Noon _____

Multiply.

1.	88	98	85	96	63
	× 15	× 77	× 65	× 44	× 40

Subtract.

2.	198	186	132	171	165
	− 89	− 77	− 78	− 119	− 75

line graph

What was the temperature at each hour ?

1 P.M. _____

2 P.M. _____

3 P.M. _____

4 P.M. _____

5 P.M. _____

6 P.M. _____

7 P.M. _____

8 P.M. _____

Temperatures for 24 Hours

| | | | | | | | | | | | | | | | | | | | | | | | |
|64°|
|62°|
|60°|
|58°|
|56°|
|54°|
|52°|
|50°|
|48°|
|46°|
|44°|
|42°|

1 2 3 4 5 6 7 8 9 10 11 M 1 2 3 4 5 6 7 8 9 10 11 N
P.M.   A.M.

Time

Temperature *(vertical axis label)*

9 P.M. _____   1 A.M. _____   5 A.M. _____   9 A.M. _____

10 P.M. _____   2 A.M. _____   6 A.M. _____   10 A.M. _____

11 P.M. _____   3 A.M. _____   7 A.M. _____   11 A.M. _____

Midnight _____   4 A.M. _____   8 A.M. _____   Noon _____

Add.

1.

```
   384        123         23        121        763
   108         38         76         34        102
    42        345        132        268         16
 + 188       + 19       +389       +328       + 23
```

2.

```
 $11.45      $7.42      $ .47       $7.04      $5.14
 +  2.15     + 4.51     + 3.81      + .55      + .09
```

Subtract.

3.

```
   81         34         65         41        158        725
 - 36       - 29       - 28       - 23       - 87       -333
```

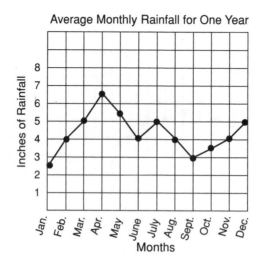

Average Monthly Rainfall for One Year

What was the average rainfall each month ?

January _____   May _____   September _____

February _____   June _____   October _____

March _____   July _____   November _____

April _____   August _____   December _____

Multiply.

1.
$$
\begin{array}{r} 89 \\ \times\ 4 \\ \hline \end{array}
\qquad
\begin{array}{r} 35 \\ \times\ 9 \\ \hline \end{array}
\qquad
\begin{array}{r} 205 \\ \times\ 6 \\ \hline \end{array}
\qquad
\begin{array}{r} 115 \\ \times\ 5 \\ \hline \end{array}
\qquad
\begin{array}{r} 431 \\ \times\ 8 \\ \hline \end{array}
$$

2.
$$
\begin{array}{r} 49 \\ \times 26 \\ \hline \end{array}
\qquad
\begin{array}{r} 37 \\ \times 59 \\ \hline \end{array}
\qquad
\begin{array}{r} 56 \\ \times 93 \\ \hline \end{array}
\qquad
\begin{array}{r} 25 \\ \times 14 \\ \hline \end{array}
\qquad
\begin{array}{r} 45 \\ \times 17 \\ \hline \end{array}
$$

Divide.

3. $3\overline{)366}$ $\qquad$ $4\overline{)440}$ $\qquad$ $5\overline{)500}$ $\qquad$ $6\overline{)18}$ $\qquad$ $8\overline{)48}$ $\qquad$ $7\overline{)63}$

Use the table to complete the graph.

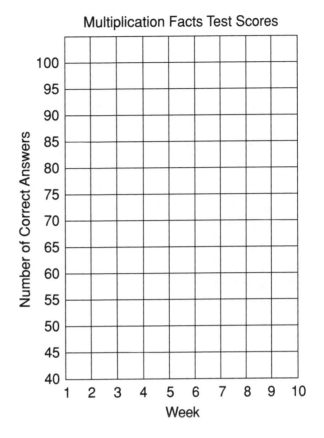

Week	Number of correct anwers
1	50
2	55
3	65
4	60
5	65
6	70
7	80
8	82
9	85
10	90

Divide and check.

1. 5) 4 2   6) 5 6   3) 3 8

2. 5) 3 2 7   4) 2 6 8   3) 8 7

3. 3) 2 7 1   5) 3 6 7   4) 2 8 4

Use the table to complete the graph.

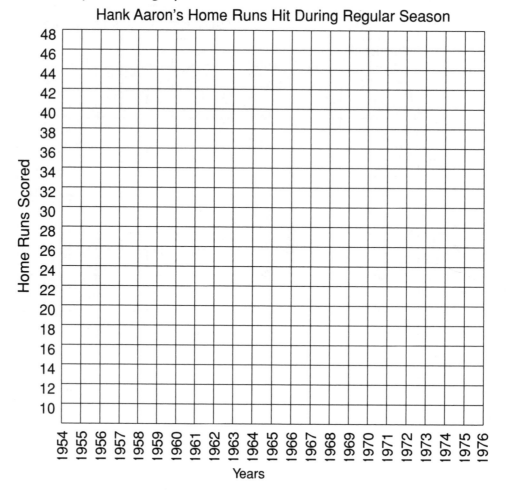

Hank Aaron's Home Runs Hit During Regular Season

Year	Number HR scored
1954	13
1955	27
1956	26
1957	44
1958	30
1959	39
1960	40
1961	34

Year	Number HR scored
1962	45
1963	44
1964	24
1965	32
1966	44
1967	39
1968	29
1969	44

Year	Number HR scored
1970	38
1971	47
1972	34
1973	40
1974	20
1975	12
1976	10

1. What year did Hank Aaron score the most home runs ? _____

2. How many more home runs did he score in 1966 than in 1967 ? _____

3. How many runs did he score his last three years put together ? _____

4. What were four great years for Hank Aaron ? _____ _____ _____ _____

5. How many home runs did he score during all 23 years ? _____

Multiply.

1.	325	716	804	439	325
	× 3	× 4	× 6	× 8	× 5

2.	174	72	15	86	41
	× 5	× 32	× 86	× 54	× 19

3.	36	59	64	37	92
	× 36	× 82	× 29	× 47	× 15

Subtract.

4.	532	607	428	628	300
	− 114	− 123	− 107	− 579	− 143

Divide.

5.	8)‾16‾	3)‾963‾	2)‾122‾	5)‾205‾	9)‾189‾

Add.

6.	29	85	88	25	79
	76	30	65	82	54
	+ 25	+ 42	+ 28	+ 71	+ 42

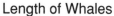

Length of Whales

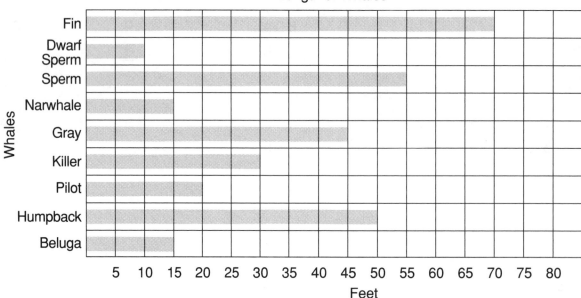

1. The longest whale: _____

2. The shortest whale: _____

3. Two whales that are about the same length: _____

4. Which whale is longer than the killer
   whale but shorter than the humpback whale ? _____

5. Which whale is longer than the pilot
   whale but shorter than the gray whale ? _____

Complete this table. List the whales in order. Put the longest first. Put the shortest last.

Rank	Whale	Length in feet

bar graph

Finish this horizontal bar graph.

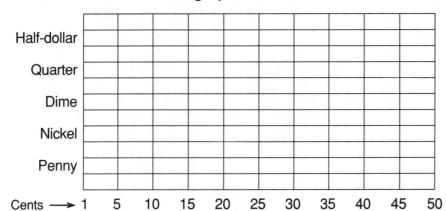

Half-dollar _____ ¢

Quarter _____ ¢

Dime _____ ¢

Nickel _____ ¢

Penny _____ ¢

You buy:	It costs:	You give the clerk:	What is your change ?
nuts	$0.98	$1.00	_____
milk	$0.69	$0.75	_____
gelatin	$0.56	$0.75	_____
soap	$0.59	$1.00	_____
cereal	$4.89	$10.00	_____

1.

```
   68      72      97      11      34      20      83      56
   40      35      10      97      75      89      25      52
   15      20      39      16      77      40      23      51
 + 62    + 78    + 60    + 91    + 32    + 47    + 84    + 58
```

2.

```
   65     100      76      84      87     109     114
 - 30    - 60    - 75    - 14    - 17    - 59    - 84
```

3.

```
  101      48      92     102     124     100
 × 81    × 20    × 90    × 62    × 90    × 30
```

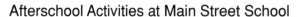

Afterschool Activities at Main Street School

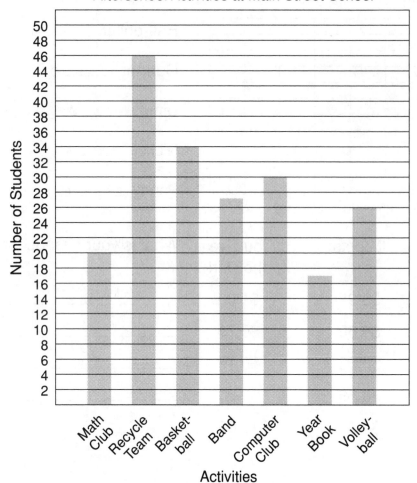

Fill in this table.

Activity	Number of Students
Math Club	
Recycle Team	
Basketball	
Band	
Computer Club	
Yearbook	
Volleyball	

Multiply.

1.
24
× 32

15
× 12

48
× 26

54
× 71

92
× 39

Divide and check.

2. 3)428

2)471

5)368

Miles Walked in Walk-A-Thon

Name	Miles Walked
Tom	
Mary	
Jose	
Tony	
Sheli	
Mike	
Kirsten	

Tom

Mary

Jose

Tony

Sheli

Mike

Kirsten

= 2 miles walked

1.

48	54	27	65	33
22	67	63	37	48
38	29	18	65	41
+ 53	+ 81	+ 84	+ 45	+ 39

2.

106	71	48	73	71
− 52	− 59	− 45	− 68	− 43

3.

46	82	46	95	46
× 37	× 23	× 21	× 83	× 80

Draw a pictograph to show the number of moons that orbit each planet.

Moons of the Planets

Planet	Number of Moons
Mercury	0
Venus	0
Earth	1
Mars	2
Jupiter	16
Saturn	18
Uranus	15
Neptune	8
Pluto	1

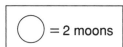 = 2 moons

Number of Moons

Mercury Venus Earth Mars Jupiter Saturn Uranus Neptune Pluto

Planets

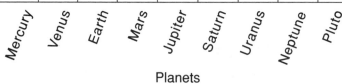

1.

```
   42        132        342        114        342
 × 26      × 24       × 24       × 34       × 34
```

2. 4)386   4)972   4)631   3)696

Wingspan of Butterflies of the World

Butterfly	Wingspan
Orange Oakleaf	3 1/2"
Karner Blue	1"
Royal Swallowtail	4"
Christmas	4 1/2"
American Lady	2"
Goliath Birdwing	11"
Julia	3"
Monarch	4 1/2"
Queen Alexandria	12"
Saturn	4"

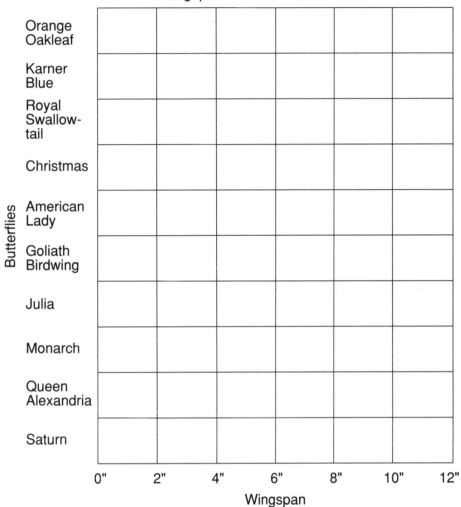

Add or subtract.

1.

41	33	45	98	70
34	40	96	73	82
94	78	28	87	84
+ 87	+ 54	+ 40	+ 30	+ 74

2.

30	62	40	82	200
− 4	− 13	− 23	− 15	− 86

Divide each shape into 4 equal parts.

 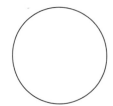 

A **fraction** has a **numerator** and a **denominator**.

$$\frac{\textbf{numerator}}{\textbf{denominator}} = \frac{\text{number of parts being considered}}{\text{number of parts that make the whole}}$$

Write a fraction to show what part of the whole is shaded.

1.  ____    ____   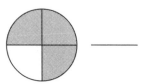 ____

2. 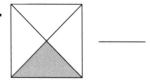 ____   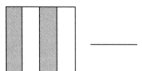 ____   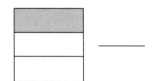 ____

3. 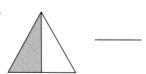 ____   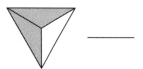 ____   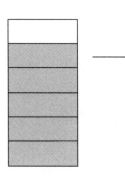 ____

fractions

 $\dfrac{2}{3} = \dfrac{\text{number of parts being considered}}{\text{number of parts that make the whole}}$

Write a fraction to show what part of the whole is shaded.

1. _____   _____    _____

2. 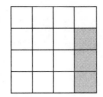 _____   _____   _____

3.  _____   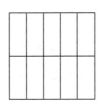 _____    _____

Shade parts of the whole to show the fraction.

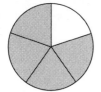 $\dfrac{1}{3}$  $\dfrac{3}{10}$ 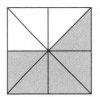 $\dfrac{3}{8}$

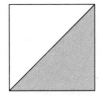 $\dfrac{5}{6}$ $\dfrac{5}{8}$ $\dfrac{1}{5}$

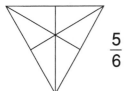 $\dfrac{1}{6}$ 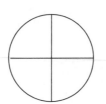 $\dfrac{4}{4}$ 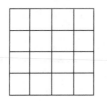 $\dfrac{7}{16}$

$\dfrac{5}{6}$ = $\dfrac{\text{number of parts being considered}}{\text{number of parts that make the whole}}$

fractions

1. Write the fraction that tells what part of the whole is shaded.

———   ———   ———

———   ———   ———

Multiply.

2.
   85
× 32
         24
× 51
         93
× 67
         20
× 48
         30
× 90

Divide.

3. 5)325   6)397   5)461   2)183

Write the fraction that tells what part of the whole is shaded.
Draw lines to connect **equivalent** fractions.

1. 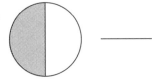 ——    ——

2. 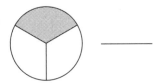 ——   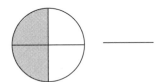 ——

3. 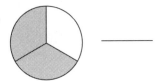 ——   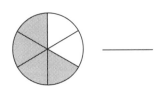 ——

4. 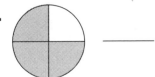 ——   ——

Add.

5.
$$
\begin{array}{r} 346 \\ +203 \end{array} \qquad
\begin{array}{r} 423 \\ +576 \end{array} \qquad
\begin{array}{r} 309 \\ +286 \end{array} \qquad
\begin{array}{r} 427 \\ +299 \end{array} \qquad
\begin{array}{r} 371 \\ +489 \end{array}
$$

6.
$$
\begin{array}{r} 434 \\ +276 \end{array} \qquad
\begin{array}{r} 827 \\ +293 \end{array} \qquad
\begin{array}{r} 461 \\ +209 \end{array} \qquad
\begin{array}{r} 371 \\ +315 \end{array} \qquad
\begin{array}{r} 462 \\ +897 \end{array}
$$

Subtract.

7.
$$
\begin{array}{r} 400 \\ -137 \end{array} \qquad
\begin{array}{r} 623 \\ -107 \end{array} \qquad
\begin{array}{r} 392 \\ -177 \end{array} \qquad
\begin{array}{r} 846 \\ -299 \end{array} \qquad
\begin{array}{r} 409 \\ -273 \end{array}
$$

8.
$$
\begin{array}{r} 800 \\ -286 \end{array} \qquad
\begin{array}{r} 403 \\ -263 \end{array} \qquad
\begin{array}{r} 523 \\ -176 \end{array} \qquad
\begin{array}{r} 827 \\ -127 \end{array} \qquad
\begin{array}{r} 342 \\ -149 \end{array}
$$

Write the **equivalent** fraction.

1.

$$\frac{1}{4} = \frac{}{8}$$

 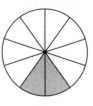

$$\frac{1}{5} = \frac{}{10}$$

2.

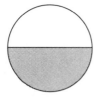 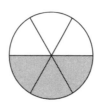

$$\frac{1}{2} = \frac{}{6}$$

 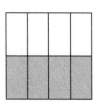

$$\frac{1}{2} = \frac{}{8}$$

Complete these factor trees.

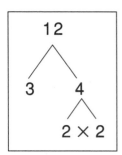

```
      12
      /\
     3  4
        /\
      2 × 2
```

3.

```
     10
     /\
    5 × 
```

```
     15
     /\
    3 × 
```

```
     35
     /\
    7 × 
```

4.

```
     18
     /\
    9 × 
```

```
     18
     /\
    3 × 
```

```
     48
     /\
    6 × 
```

Multiply.

5.

$$\begin{array}{r} 36 \\ \times\,25 \\ \hline \end{array} \qquad \begin{array}{r} 52 \\ \times\,38 \\ \hline \end{array} \qquad \begin{array}{r} 47 \\ \times\,53 \\ \hline \end{array} \qquad \begin{array}{r} 86 \\ \times\,21 \\ \hline \end{array} \qquad \begin{array}{r} 40 \\ \times\,37 \\ \hline \end{array}$$

Write the equivalent fraction.

1. $\dfrac{3}{4} = \dfrac{}{8}$ $\dfrac{2}{3} = \dfrac{}{6}$ $\dfrac{1}{2} = \dfrac{}{8}$

2. $\dfrac{2}{3} = \dfrac{}{9}$ $\dfrac{1}{2} = \dfrac{}{6}$ $\dfrac{2}{5} = \dfrac{}{10}$

3. $\dfrac{3}{4} = \dfrac{}{12}$ $\dfrac{2}{6} = \dfrac{}{12}$ $\dfrac{1}{3} = \dfrac{}{9}$

Complete these factor trees. Circle the common factors.

4.

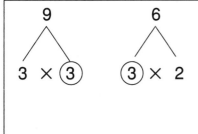

9   6   3   6

3 × ③   ③ × 2

5. 6   24   5   25

6. 8   20   7   21

Write the equivalent fraction.

    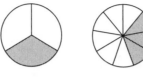   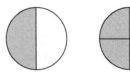

1. $\dfrac{1}{3} = \dfrac{}{}$ $\dfrac{1}{3} = \dfrac{}{9}$ $\dfrac{1}{2} = \dfrac{}{4}$

2. $\dfrac{2}{3} = \dfrac{}{15}$ $\dfrac{4}{5} = \dfrac{}{10}$ $\dfrac{1}{2} = \dfrac{}{8}$

3. $\dfrac{3}{8} = \dfrac{}{16}$ $\dfrac{4}{8} = \dfrac{}{16}$ $\dfrac{3}{4} = \dfrac{}{12}$

4. Complete the factor trees. Circle the common factors.
Find the **greatest common factor** (GCF).

12   12   12   18   18   18

12   1   ⑥   2   4   3   18   1   9   2   ⑥   3

GCF = __6__

14   14   12   12   12

GCF = ___

Write the equivalent fractions.

1. $\dfrac{2}{5} = \dfrac{}{15}$ $\dfrac{3}{7} = \dfrac{}{14}$ $\dfrac{2}{10} = \dfrac{}{20}$

2. $\dfrac{1}{4} = \dfrac{}{16}$ $\dfrac{2}{3} = \dfrac{}{12}$ $\dfrac{3}{5} = \dfrac{}{10}$

Complete these factor trees. Find the greatest common factors.

3.   14   12   10   15

GCF = ____   GCF = ____

Multiply.

4. $\begin{array}{r} 89 \\ \times\ 56 \\ \hline \end{array}$ $\begin{array}{r} 75 \\ \times\ 19 \\ \hline \end{array}$ $\begin{array}{r} 42 \\ \times\ 61 \\ \hline \end{array}$ $\begin{array}{r} 17 \\ \times\ 23 \\ \hline \end{array}$

Divide.

5. $6\overline{)187}$ $4\overline{)276}$ $3\overline{)143}$ $5\overline{)379}$

---

simplify fractions

Circle the fraction that is in
lowest terms.

Write the fraction that shows
what part is shaded.

1.
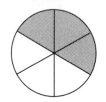
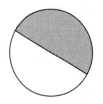

$$\frac{3}{6} \qquad\qquad \frac{1}{2}$$

_____   _____

Find the greatest common factors (GCF).

2.   4   12   16   18

GCF = ___   GCF = ___

3.   6   10   5   15

GCF = ___   GCF = ___

Use the GCF to simplify these fractions to lowest terms.

4. $\dfrac{4}{12} =$ $\dfrac{16}{18} =$ $\dfrac{6}{9} =$

5. $\dfrac{6}{10} =$ $\dfrac{5}{15} =$ $\dfrac{9}{18} =$

126

Measure these line segments. Write the measurements on the line segment.

1. └──────────────┘   └──────────────┐

2. └────────────────────┘   └─────────────────────┐

3. └────────────────────────────┘

Simplify the fractions to lowest terms.

4. $\frac{4}{8}$ = $\frac{3}{6}$ = $\frac{6}{8}$ =

5. $\frac{3}{9}$ = $\frac{6}{9}$ = $\frac{8}{10}$ =

6. $\frac{2}{4}$ = $\frac{5}{10}$ = $\frac{3}{12}$ =

Add.

7.
$$
\begin{array}{r} 362 \\ + 276 \\ \hline \end{array}
\qquad
\begin{array}{r} 427 \\ + 376 \\ \hline \end{array}
\qquad
\begin{array}{r} 823 \\ + 409 \\ \hline \end{array}
\qquad
\begin{array}{r} 703 \\ + 317 \\ \hline \end{array}
\qquad
\begin{array}{r} 429 \\ + 286 \\ \hline \end{array}
$$

Subtract.

8.
$$
\begin{array}{r} 872 \\ - 461 \\ \hline \end{array}
\qquad
\begin{array}{r} 304 \\ - 172 \\ \hline \end{array}
\qquad
\begin{array}{r} 864 \\ - 159 \\ \hline \end{array}
\qquad
\begin{array}{r} 372 \\ - 168 \\ \hline \end{array}
\qquad
\begin{array}{r} 203 \\ - 197 \\ \hline \end{array}
$$

Write equivalent fractions.

1. $\dfrac{2}{3} = \dfrac{}{12}$ $\dfrac{3}{5} = \dfrac{}{15}$ $\dfrac{4}{6} = \dfrac{}{12}$

2. $\dfrac{3}{6} = \dfrac{}{18}$ $\dfrac{2}{4} = \dfrac{}{16}$ $\dfrac{3}{5} = \dfrac{}{40}$

3. $\dfrac{1}{2} = \dfrac{}{12}$ $\dfrac{2}{3} = \dfrac{}{9}$ $\dfrac{3}{4} = \dfrac{}{16}$

Simplify these fractions to lowest terms.

4. $\dfrac{6}{8} =$ $\dfrac{5}{10} =$ $\dfrac{3}{9} =$

5. $\dfrac{8}{12} =$ $\dfrac{9}{18} =$ $\dfrac{3}{6} =$

Draw line segments that measure:

$3\dfrac{1}{2}$ inches

$1\dfrac{3}{4}$ inches $1\dfrac{1}{4}$ inches

Shade in to show fractional parts.

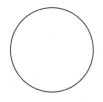  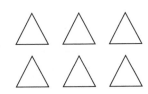  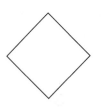

$\dfrac{3}{8}$ $\qquad$ $\dfrac{1}{5}$ $\qquad$ $\dfrac{5}{6}$ $\qquad$ $\dfrac{2}{3}$ $\qquad$ $\dfrac{3}{4}$

Complete these factor trees. Circle the common factors.

1.   12   14   6   12

GCF = ___   GCF = ___

Simplify the fractions to lowest terms.

2. $\dfrac{5}{10} =$ $\dfrac{4}{6} =$ $\dfrac{8}{12} =$

3. $\dfrac{6}{9} =$ $\dfrac{2}{4} =$ $\dfrac{12}{20} =$

Write equivalent fractions.

4. $\dfrac{4}{5} = \dfrac{}{10}$ $\dfrac{3}{7} = \dfrac{}{14}$ $\dfrac{2}{3} = \dfrac{}{12}$

5. $\dfrac{3}{8} = \dfrac{}{16}$ $\dfrac{2}{9} = \dfrac{}{18}$ $\dfrac{3}{6} = \dfrac{}{18}$

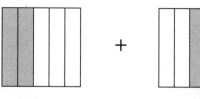

$\dfrac{2}{5}$ + $\dfrac{1}{5}$ = $\dfrac{3}{5}$

$\begin{array}{r} \dfrac{2}{5} \\[2mm] +\dfrac{1}{5} \\[1mm] \hline \dfrac{3}{5} \end{array}$

Add.

1. $\dfrac{6}{8} + \dfrac{1}{8} = \dfrac{}{8}$ 　　　 $\dfrac{2}{5} + \dfrac{2}{5} = \dfrac{}{5}$ 　　　 $\dfrac{2}{9} + \dfrac{3}{9} = \dfrac{}{9}$

2. $\dfrac{5}{7} + \dfrac{1}{7} = \underline{}$ 　　　 $\dfrac{3}{8} + \dfrac{2}{8} = \underline{}$ 　　　 $\dfrac{1}{3} + \dfrac{1}{3} = \underline{}$

3.
$\begin{array}{r} \dfrac{3}{5} \\[2mm] +\dfrac{1}{5} \\[1mm] \hline \end{array}$ 　　　
$\begin{array}{r} \dfrac{4}{7} \\[2mm] +\dfrac{2}{7} \\[1mm] \hline \end{array}$ 　　　
$\begin{array}{r} \dfrac{7}{12} \\[2mm] +\dfrac{4}{12} \\[1mm] \hline \end{array}$ 　　　
$\begin{array}{r} \dfrac{4}{9} \\[2mm] +\dfrac{4}{9} \\[1mm] \hline \end{array}$ 　　　
$\begin{array}{r} \dfrac{3}{10} \\[2mm] +\dfrac{4}{10} \\[1mm] \hline \end{array}$

4.
$\begin{array}{r} \dfrac{2}{7} \\[2mm] +\dfrac{3}{7} \\[1mm] \hline \end{array}$ 　　　
$\begin{array}{r} \dfrac{3}{12} \\[2mm] +\dfrac{2}{12} \\[1mm] \hline \end{array}$ 　　　
$\begin{array}{r} \dfrac{2}{6} \\[2mm] +\dfrac{3}{6} \\[1mm] \hline \end{array}$ 　　　
$\begin{array}{r} \dfrac{1}{7} \\[2mm] +\dfrac{3}{7} \\[1mm] \hline \end{array}$ 　　　
$\begin{array}{r} \dfrac{3}{8} \\[2mm] +\dfrac{4}{8} \\[1mm] \hline \end{array}$

5. Tom ate $\dfrac{3}{8}$ of the pizza.

Bill ate $\dfrac{2}{8}$ of the pizza.

How much did they eat together ? _____

Use the diagram to show your work.

$$\frac{2}{3}$$
$$-\frac{1}{3}$$
$$\frac{1}{3}$$

$$\frac{5}{6}$$
$$-\frac{4}{6}$$
$$\frac{1}{6}$$

Subtract.

1.
$$\frac{5}{12}$$
$$-\frac{4}{12}$$

$$\frac{5}{9}$$
$$-\frac{3}{9}$$

$$\frac{4}{5}$$
$$-\frac{2}{5}$$

2. $\frac{3}{5} - \frac{1}{5} = \frac{\quad}{5}$

$\frac{4}{5} - \frac{3}{5} = \frac{\quad}{5}$

$\frac{6}{7} - \frac{2}{7} = \frac{\quad}{7}$

3. $\frac{7}{8} - \frac{2}{8} = \frac{\quad}{8}$

$\frac{5}{8} - \frac{2}{8} = \underline{\quad}$

$\frac{5}{6} - \frac{4}{6} = \underline{\quad}$

4. $\frac{15}{16} - \frac{6}{16} = \underline{\quad}$

$\frac{7}{8} - \frac{7}{8} = \underline{\quad}$

$\frac{11}{14} - \frac{8}{14} = \underline{\quad}$

5. Lori had $\frac{4}{5}$ of a can of

paint. She used $\frac{1}{5}$ of the

paint. How much is left ? _____

6. You have $\frac{5}{6}$ of a yard of cloth.

You use $\frac{4}{6}$ of the yard.

How much cloth is left ? _____

$\frac{1}{6}$	$\frac{1}{6}$	$\frac{1}{6}$	$\frac{1}{6}$	$\frac{1}{6}$

Write these improper fractions as mixed numbers.

$$\frac{1}{4} \quad = \quad 4\overline{)11} \begin{array}{l} 2\frac{3}{4} \\ \hline 11 \\ -\,8 \\ \hline 3 \end{array}$$

$$\frac{13}{5} \quad = \quad 5\overline{)13} \begin{array}{l} 2\frac{3}{5} \\ \hline 13 \\ -\,10 \\ \hline 3 \end{array}$$

1. $\frac{15}{4} \quad = \quad 4\overline{)15}$ $\qquad \frac{9}{8} \quad = \quad 8\overline{)9}$ $\qquad \frac{1}{6} \quad = \quad 6\overline{)11}$

2. $\frac{1}{8} \quad =$ $\qquad \frac{18}{12} \quad =$ $\qquad \frac{12}{7} \quad =$

3. $\frac{13}{10} \quad =$ $\qquad \frac{16}{7} \quad =$ $\qquad \frac{1}{10} \quad =$

Add. Simplify improper fractions.

4. $\begin{array}{r} \frac{4}{9} \\ +\,\frac{7}{9} \\ \hline \end{array}$ $\qquad \begin{array}{r} \frac{7}{10} \\ +\,\frac{6}{10} \\ \hline \end{array}$ $\qquad \begin{array}{r} \frac{6}{9} \\ +\,\frac{5}{9} \\ \hline \end{array}$ $\qquad \begin{array}{r} \frac{3}{5} \\ +\,\frac{3}{5} \\ \hline \end{array}$ $\qquad \begin{array}{r} \frac{5}{7} \\ +\,\frac{6}{7} \\ \hline \end{array}$

Write the mixed number as an improper fraction.

$$2 \overset{+1}{\underset{\times 4}{}} = \frac{9}{4}$$

$$3 \overset{+2}{\underset{\times 3}{}} = \frac{11}{3}$$

1. $2\frac{1}{2} = \frac{}{2}$　　　　$3\frac{1}{8} = \frac{}{8}$　　　　$2\frac{2}{5} = \frac{}{5}$

2. $3\frac{3}{5} = \frac{}{5}$　　　　$4\frac{3}{4} = \frac{}{4}$　　　　$5\frac{1}{8} = \frac{}{8}$

3. $5\frac{2}{5} = \frac{}{5}$　　　　$1\frac{5}{6} = \frac{}{6}$　　　　$2\frac{2}{3} = \frac{}{3}$

4. $6\frac{1}{2} = \frac{}{2}$　　　　$5\frac{1}{3} = \frac{}{3}$　　　　$7\frac{2}{5} = \frac{}{5}$

Multiply.

5.　　$\begin{array}{r} 32 \\ \times\,27 \\ \hline \end{array}$　　　　$\begin{array}{r} 46 \\ \times\,58 \\ \hline \end{array}$　　　　$\begin{array}{r} 59 \\ \times\,36 \\ \hline \end{array}$　　　　$\begin{array}{r} 27 \\ \times\,15 \\ \hline \end{array}$

Divide.

6.　　$5\overline{)235}$　　　　$6\overline{)386}$　　　　$7\overline{)174}$　　　　$8\overline{)621}$

| **a** | *fractions* |

Write the mixed numbers as improper fractions.

1. $2\frac{1}{2} = \frac{}{2}$ $3\frac{3}{4} = \frac{}{4}$ $2\frac{1}{5} = \frac{}{5}$ $6\frac{1}{8} = \frac{}{8}$

2. $3\frac{3}{5} = \frac{}{5}$ $2\frac{1}{7} = \frac{}{7}$ $8\frac{2}{3} = \frac{}{3}$ $4\frac{1}{2} = \frac{}{2}$

Write improper fractions as mixed numbers.

3. $\frac{8}{3} =$ $\frac{12}{7} =$ $\frac{16}{3} =$ $\frac{11}{4} =$

4. $\frac{9}{7} =$ $\frac{5}{4} =$ $\frac{7}{3} =$ $\frac{5}{2} =$

Simplify these fractions.

5. $\frac{6}{8} =$ $\frac{2}{4} =$ $\frac{3}{6} =$ $\frac{8}{12} =$

6. $\frac{4}{6} =$ $\frac{6}{9} =$ $\frac{8}{10} =$ $\frac{5}{15} =$

Add or subtract fractions. Simplify the answers.

7.
$$\begin{array}{r} \frac{3}{4} \\ +\frac{1}{4} \\ \hline \end{array} \qquad \begin{array}{r} \frac{4}{9} \\ +\frac{2}{9} \\ \hline \end{array} \qquad \begin{array}{r} \frac{5}{10} \\ +\frac{3}{10} \\ \hline \end{array} \qquad \begin{array}{r} \frac{11}{12} \\ +\frac{6}{12} \\ \hline \end{array} \qquad \begin{array}{r} \frac{3}{15} \\ +\frac{12}{15} \\ \hline \end{array}$$

8.
$$\begin{array}{r} \frac{5}{6} \\ -\frac{1}{6} \\ \hline \end{array} \qquad \begin{array}{r} \frac{5}{6} \\ -\frac{2}{6} \\ \hline \end{array} \qquad \begin{array}{r} \frac{8}{9} \\ -\frac{2}{9} \\ \hline \end{array} \qquad \begin{array}{r} \frac{3}{4} \\ -\frac{1}{4} \\ \hline \end{array} \qquad \begin{array}{r} \frac{8}{9} \\ -\frac{2}{9} \\ \hline \end{array}$$

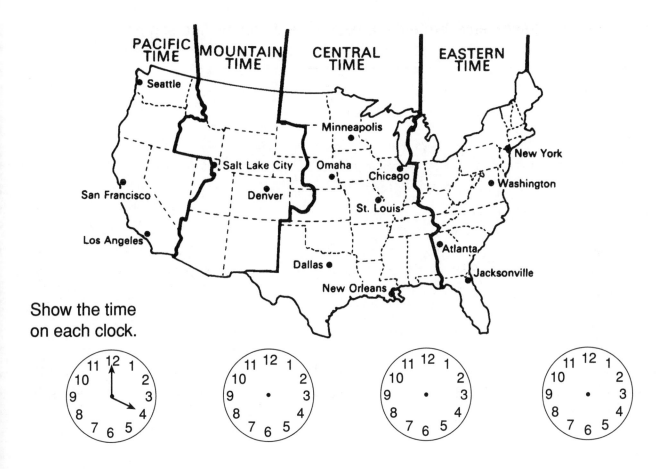

Show the time
on each clock.

Add and subtract fractions. Simplify the answers.

1.

$\frac{3}{8}$ $\quad$ $\frac{4}{5}$ $\quad$ $\frac{3}{8}$ $\quad$ $\frac{4}{12}$ $\quad$ $\frac{9}{12}$

$+\frac{6}{8}$ $\quad$ $+\frac{2}{5}$ $\quad$ $+\frac{3}{8}$ $\quad$ $+\frac{4}{12}$ $\quad$ $+\frac{4}{12}$

2.

$\frac{5}{6}$ $\quad$ $\frac{7}{8}$ $\quad$ $\frac{9}{9}$ $\quad$ $\frac{8}{12}$ $\quad$ $\frac{9}{10}$

$-\frac{1}{6}$ $\quad$ $-\frac{3}{8}$ $\quad$ $-\frac{3}{9}$ $\quad$ $-\frac{2}{12}$ $\quad$ $-\frac{6}{10}$

3.

$\frac{9}{10}$ $\quad$ $\frac{8}{14}$ $\quad$ $\frac{7}{9}$ $\quad$ $\frac{9}{10}$ $\quad$ $\frac{6}{15}$

$-\frac{4}{10}$ $\quad$ $-\frac{1}{14}$ $\quad$ $-\frac{4}{9}$ $\quad$ $-\frac{4}{10}$ $\quad$ $-\frac{3}{15}$

Measure these line segments. Write the measurements on the line segments.

Add.

1.

42	37	34	83	72
53	89	+ 42	+ 21	+ 99
+ 64	+ 42			

2.

381	407	392	464	823
+ 217	+ 283	+ 209	+ 357	+ 799

Subtract.

3.

28	67	249	460	601
− 14	− 38	− 187	− 238	− 598

Multiply.

4.

37	406	38	80	97
× 5	× 7	× 42	× 58	× 68

Divide.

5. $5\overline{)18}$ $6\overline{)42}$ $7\overline{)368}$ $8\overline{)428}$ $9\overline{)847}$

136

Complete the multiples.

1. multiples of 2: 2, 4, 6, 8, 10, 12, 14, 16, _____ , _____ .

2. multiples of 3: 3, 6, 9, 12, _____ , _____ , _____ , _____ , _____ , _____ .

3. multiples of 4: 8, 12, _____ , _____ , _____ , _____ , _____ , _____ , _____ .

4. multiples of 6: 6, 12, 18, _____ , _____ , _____ , _____ , _____ , _____ , _____ .

5. multiples of 7: 7, 14, _____ , _____ , _____ , _____ , _____ , _____ , _____ .

6. multiples of 8: 8, 16, _____ , _____ , _____ , _____ , _____ , _____ , _____ .

7. multiples of 9: 9, 18, _____ , _____ , _____ , _____ , _____ , _____ , _____ .

Common multiples of 2 and 3:

Multiples of 2: 2, 4, 6, 8, 10, 12, 14, 16, 18, 20

Multiples of 3: 3, 6, 9, 8, 12, 15, 18, 21, 24, 27, 30

Least common multiple of 2 and 3 = 6

Common multiples of 6 and 8:

8.   multiples of 6: _____ , _____ , _____ , _____ , _____ , _____ , _____ .

9.   multiples of 8: _____ , _____ , _____ , _____ , _____ , _____ , _____ .

10.   Least common multiple of 6 and 8 = _____ .

Write the LCM (least common multiple).

11. 2
        6
            LCM = _____

12. 3
        8
            LCM = _____

13. 4
        5
            LCM = _____

14. 6
        9
            LCM = _____

---

1. Complete the multiples.

6: 6, 12, ____ , ____ , ____ , ____ , ____ , ____ , ____ , ____ .

7: 7, 14, ____ , ____ , ____ , ____ , ____ , ____ , ____ , ____ .

Circle the common multiples.

LCM = ____

Find the lowest common multiples.

2. 7

   5   LCM = ____

3. 8

   10   LCM = ____

4. 4

   6   LCM = ____

5. 2

   9   LCM = ____

Add fractions. Simplity the answers.

6.

$$\frac{1}{10} + \frac{4}{10}$$

$$\frac{1}{12} + \frac{1}{12}$$

$$\frac{4}{8} + \frac{5}{8}$$

$$\frac{5}{8} + \frac{3}{8}$$

$$\frac{3}{4} + \frac{2}{4}$$

Subtract fractions. Simply the answers.

7.

$$\frac{5}{16} - \frac{1}{16}$$

$$\frac{8}{9} - \frac{5}{9}$$

$$\frac{9}{9} - \frac{6}{9}$$

$$\frac{6}{10} - \frac{2}{10}$$

$$\frac{8}{12} - \frac{2}{12}$$

Find the lowest common multiples.

1. 4
8   LCM = ____

6
9   LCM = ____

2. 2
4   LCM = ____

4
6   LCM = ____

Write the equivalent fractions.

3. $\dfrac{2}{3} = \dfrac{}{6}$ $\dfrac{3}{5} = \dfrac{}{20}$ $\dfrac{3}{8} = \dfrac{}{24}$ $\dfrac{2}{8} = \dfrac{}{16}$

4. $\dfrac{1}{2} = \dfrac{}{8}$ $\dfrac{1}{4} = \dfrac{}{16}$ $\dfrac{2}{3} = \dfrac{}{12}$ $\dfrac{3}{5} = \dfrac{}{15}$

Add fractions. Simplify the answers.

5. $\dfrac{1}{2} = \dfrac{}{4}$ $\dfrac{1}{2} = \dfrac{}{10}$ $\dfrac{3}{4} = \dfrac{}{8}$

$+\dfrac{1}{4} = \dfrac{1}{4}$ $+\dfrac{1}{5} = \dfrac{}{10}$ $+\dfrac{1}{8} = \dfrac{}{8}$
_____   _____   _____

6. $\dfrac{3}{8} = \dfrac{}{}$ $\dfrac{1}{6} = \dfrac{}{10}$ $\dfrac{1}{2} = \dfrac{}{8}$

$+\dfrac{1}{2} = \dfrac{}{}$ $+\dfrac{1}{3} = \dfrac{}{10}$ $+\dfrac{1}{8} = \dfrac{}{8}$
_____   _____   _____

adding/subtracting unlike fractions

Add fractions. Simplify the answers.

1.

$\dfrac{1}{2} = \underline{}$ $\dfrac{3}{4} = \dfrac{}{8}$ $\dfrac{2}{3} = \dfrac{}{9}$

$+\dfrac{1}{3} = \dfrac{1}{6}$ $+\dfrac{3}{8} = \dfrac{3}{8}$ $+\dfrac{1}{9} = \dfrac{1}{9}$

2.

$\dfrac{4}{12} = \dfrac{}{12}$ $\dfrac{1}{8} = \dfrac{}{8}$ $\dfrac{1}{4} = \dfrac{}{16}$

$+\dfrac{1}{3} = \dfrac{}{12}$ $+\dfrac{1}{4} = \dfrac{}{8}$ $+\dfrac{3}{16} = \dfrac{}{16}$

Subtract fractions. Simplify the answers.

3.

$\dfrac{3}{8} = \underline{}$ $\dfrac{3}{8} = \underline{}$ $\dfrac{4}{5} = \underline{}$

$-\dfrac{1}{16} = \dfrac{}{16}$ $-\dfrac{1}{4} = \underline{}$ $-\dfrac{1}{3} = \underline{}$

4.

$\dfrac{2}{3} = \underline{}$ $\dfrac{1}{2} = \underline{}$ $\dfrac{8}{9} = \underline{}$

$-\dfrac{1}{12} = \underline{}$ $-\dfrac{1}{6} = \underline{}$ $-\dfrac{2}{3} = \underline{}$

What do you want to do when you get out of school?
The 20 students in one class answered that question.
Fill in the chart with data from the graph.

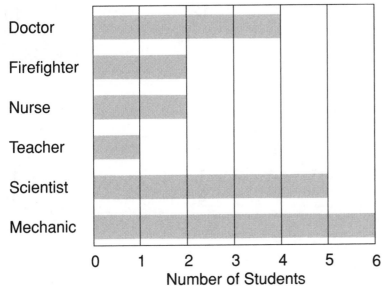

Career Goals

Career Choice	Number of students	Fractional Part of Class
Doctor	4	$\frac{4}{20} = \frac{1}{5}$
Firefighter		
Nurse		
Teacher		
Scientist		
Mechanic		

Add fractions. Simplify the answers.

1.

$$\frac{1}{2} = \frac{}{6}$$
$$+\frac{1}{3} = \frac{}{6}$$
$$\overline{\hspace{3cm}}$$

$$\frac{2}{5} = \underline{\ \ }$$
$$+\frac{1}{2} = \underline{\ \ }$$
$$\overline{\hspace{3cm}}$$

$$\frac{1}{4} = \underline{\ \ }$$
$$+\frac{3}{8} = \underline{\ \ }$$
$$\overline{\hspace{3cm}}$$

Subtract fractions. Simplify the answers.

2.

$$\frac{5}{8} = \underline{\ \ }$$
$$-\frac{1}{2} = \underline{\ \ }$$
$$\overline{\hspace{3cm}}$$

$$\frac{2}{3} = \underline{\ \ }$$
$$-\frac{1}{6} = \underline{\ \ }$$
$$\overline{\hspace{3cm}}$$

$$\frac{7}{12} = \underline{\ \ }$$
$$-\frac{1}{6} = \underline{\ \ }$$
$$\overline{\hspace{3cm}}$$

Add fractions. Simplify the answers.

1. $\dfrac{3}{8}$ $+\dfrac{6}{8}$ $\qquad$ $\dfrac{4}{5}$ $+\dfrac{2}{5}$ $\qquad$ $\dfrac{4}{12}$ $+\dfrac{9}{12}$ $\qquad$ $\dfrac{3}{10}$ $+\dfrac{8}{10}$

2. $\dfrac{1}{2}$ $+\dfrac{3}{4}$ $\qquad$ $\dfrac{3}{4}$ $+\dfrac{1}{8}$ $\qquad$ $\dfrac{3}{5}$ $+\dfrac{1}{2}$ $\qquad$ $\dfrac{7}{8}$ $+\dfrac{1}{4}$

Subtract fractions. Simplify the answers.

3. $\dfrac{3}{4}$ $-\dfrac{1}{4}$ $\qquad$ $\dfrac{3}{8}$ $-\dfrac{1}{8}$ $\qquad$ $\dfrac{3}{5}$ $-\dfrac{1}{5}$ $\qquad$ $\dfrac{5}{8}$ $-\dfrac{1}{8}$

4. $\dfrac{5}{12}$ $-\dfrac{1}{6}$ $\qquad$ $\dfrac{7}{10}$ $-\dfrac{1}{5}$ $\qquad$ $\dfrac{1}{2}$ $-\dfrac{1}{3}$ $\qquad$ $\dfrac{2}{3}$ $-\dfrac{1}{9}$

Weekend Walk-a-Thon

Walker	Saturday	Sunday
Meredith	$1\frac{1}{3}$ mile	$4\frac{1}{3}$ mile
Abe	$2\frac{3}{8}$ mile	$2\frac{2}{8}$ mile
Darcy	$3\frac{1}{4}$ mile	$1\frac{1}{4}$ mile
Roger	$4\frac{1}{10}$ mile	$1\frac{3}{10}$ mile
Susan	$1\frac{1}{8}$ mile	$3\frac{3}{8}$ mile

1. Who walked the farthest on Saturday ? _____

2. Who walked the farthest on Sunday ? _____

How far did each walker walk during the weekend ?

3. Meredith

$1\frac{1}{3}$ miles

$+2\frac{1}{3}$ miles
_____

$3\frac{2}{3}$ miles

Abe

$2\frac{3}{8}$ miles

$+1\frac{2}{8}$ miles
_____

miles

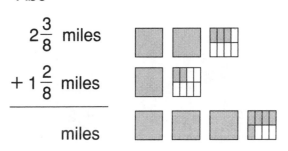

4.   Darcy   Roger   Susan

Add or subtract fractions.

5. $\dfrac{2}{3}$ $\dfrac{3}{4}$ $\dfrac{1}{4}$ $\dfrac{1}{2}$

$+\dfrac{1}{3}$ $-\dfrac{1}{2}$ $+\dfrac{1}{8}$ $-\dfrac{3}{10}$
_____   _____   _____   _____

$2\dfrac{2}{5}$

$+\ 1\dfrac{1}{5}$

$1\dfrac{1}{4} = 1\dfrac{1}{4}$

$+\ 1\dfrac{1}{2} = 1\dfrac{2}{4}$

$2\dfrac{3}{4}$

Add. Simplify your answers.

1.

$1\dfrac{1}{3}$
$+\ 2\dfrac{1}{3}$

$4\dfrac{1}{8}$
$+\ 3\dfrac{3}{8}$

$6\dfrac{1}{6}$
$+\ 4\dfrac{1}{6}$

$1\dfrac{3}{8}$
$+\ 2\dfrac{3}{8}$

2.

$4\dfrac{1}{3}$
$+\ 2\dfrac{1}{6}$

$1\dfrac{1}{2}$
$+\ 2\dfrac{1}{6}$

$3\dfrac{1}{3}$
$+\ 2\dfrac{1}{2}$

$1\dfrac{1}{4}$
$+\ 1\dfrac{3}{8}$

3.

$3\dfrac{1}{4}$
$+\ 2\dfrac{1}{4}$

$8\dfrac{1}{7}$
$+\ 3\dfrac{1}{2}$

$2\dfrac{3}{4}$
$+\ 3\dfrac{1}{8}$

$6\dfrac{2}{3}$
$+\ 5\dfrac{1}{6}$

4.

$2\dfrac{1}{2}$
$+\ 2\dfrac{1}{4}$

$3\dfrac{2}{5}$
$+\ 1\dfrac{1}{2}$

$5\dfrac{1}{2}$
$+\ 5\dfrac{1}{2}$

$1\dfrac{2}{3}$
$+\ 4\dfrac{1}{2}$

$2\dfrac{3}{4}$

$-1\dfrac{1}{4}$

$1\dfrac{2}{4} = 1\dfrac{1}{2}$

$2\dfrac{1}{2} = 2\dfrac{3}{4}$

$-1\dfrac{1}{4} = 1\dfrac{1}{4}$

$1\dfrac{1}{4}$

Subtract. Simplify your answers.

1.

$8\dfrac{3}{4}$
$-1\dfrac{1}{4}$

$9\dfrac{7}{8}$
$-2\dfrac{3}{8}$

$6\dfrac{1}{2}$
$-2\dfrac{1}{2}$

$2\dfrac{5}{8}$
$-1\dfrac{2}{8}$

2.

$3\dfrac{2}{3}$
$-1\dfrac{1}{2}$

$7\dfrac{5}{8}$
$-1\dfrac{1}{4}$

$3\dfrac{4}{5}$
$-1\dfrac{1}{2}$

$7\dfrac{3}{5}$
$-2\dfrac{1}{10}$

3.

$3\dfrac{7}{8}$
$-2\dfrac{1}{2}$

$7\dfrac{3}{5}$
$-2\dfrac{1}{3}$

$2\dfrac{3}{4}$
$-1\dfrac{1}{3}$

$4\dfrac{1}{4}$
$-2\dfrac{1}{8}$

4.

$2\dfrac{2}{3}$
$-1\dfrac{1}{4}$

$2\dfrac{1}{2}$
$-1\dfrac{1}{6}$

$12\dfrac{3}{4}$
$-1\dfrac{1}{2}$

$8\dfrac{5}{6}$
$-2\dfrac{1}{3}$

Add. Simplify your answers.

1.
$$\frac{1}{5}$$
$$+\frac{3}{5}$$

$$\frac{2}{7}$$
$$+\frac{5}{7}$$

$$\frac{3}{8}$$
$$+\frac{1}{8}$$

$$\frac{7}{10}$$
$$+\frac{1}{10}$$

2.
$$\frac{2}{3}$$
$$+\frac{1}{2}$$

$$\frac{3}{4}$$
$$+\frac{1}{8}$$

$$\frac{7}{8}$$
$$+\frac{3}{4}$$

$$\frac{3}{4}$$
$$+\frac{2}{3}$$

3.
$$2\frac{1}{4}$$
$$+3\frac{1}{4}$$

$$3\frac{2}{3}$$
$$+1\frac{1}{6}$$

$$7\frac{1}{2}$$
$$+3\frac{1}{8}$$

$$7\frac{1}{3}$$
$$+3\frac{1}{4}$$

Subtract. Simplify your answers.

4.
$$\frac{2}{3}$$
$$-\frac{1}{3}$$

$$\frac{5}{9}$$
$$-\frac{2}{9}$$

$$\frac{5}{8}$$
$$-\frac{2}{8}$$

$$\frac{7}{9}$$
$$-\frac{1}{9}$$

5.
$$\frac{2}{3}$$
$$-\frac{1}{2}$$

$$\frac{1}{2}$$
$$-\frac{1}{4}$$

$$4\frac{7}{8}$$
$$-2\frac{1}{4}$$

$$8\frac{5}{6}$$
$$-2\frac{1}{3}$$

Solve these word problems. Use information from the chart.

Hours of Television Viewing

Name	Week 1	Week 2	Week 3	Week 4
Al	$6\frac{1}{2}$	$8\frac{1}{2}$	$4\frac{3}{4}$	$7\frac{1}{2}$
Beth	$3\frac{1}{4}$	$10\frac{3}{4}$	11	$3\frac{1}{2}$
John	$7\frac{3}{4}$	$8\frac{1}{2}$	$10\frac{3}{4}$	$9\frac{1}{4}$
Susan	$12\frac{1}{2}$	$6\frac{1}{4}$	$2\frac{1}{2}$	$8\frac{3}{4}$

1. How much more TV did Al watch during week 2 than he watched during week 1 ? _____

2. How much TV did Beth watch during week 3 and week 4 together ? _____

3. How much TV did John watch during week 2 and week 4 together ? _____

4. How much more TV did Susan watch week 1 than she watched week 2 ? _____

5. How much more TV did Al watch than Beth during week 4 ? _____

Add.

1. $\$3.27$ $\$6.20$ $\$4.84$ $\$3.26$ $\$2.87$
    $+\ 4.36$ $+\ 1.03$ $+\ 1.97$ $+\ 4.95$ $+\ 3.94$

2. $\dfrac{2}{3}$ $\dfrac{1}{2}$ $\dfrac{3}{10}$ $\dfrac{2}{3}$

    $+\dfrac{5}{6}$ $+\dfrac{5}{8}$ $+\dfrac{4}{5}$ $+\dfrac{3}{4}$

3. $2\dfrac{1}{2}$ $3\dfrac{2}{3}$ $6\dfrac{1}{5}$ $4\dfrac{1}{3}$

    $+\ 1\dfrac{1}{4}$ $+\ 2\dfrac{1}{4}$ $+\ 2\dfrac{3}{10}$ $+\ 3\dfrac{1}{6}$

Subtract.

4. $\$4.23$ $\$2.00$ $\$6.21$ $\$4.19$ $\$6.66$
    $-\ 1.12$ $-\ 1.53$ $-\ 1.50$ $-\ 2.25$ $-\ 1.89$

5. $\dfrac{5}{6}$ $\dfrac{7}{8}$ $\dfrac{3}{4}$ $\dfrac{2}{3}$

    $-\dfrac{1}{3}$ $-\dfrac{1}{4}$ $-\dfrac{1}{2}$ $-\dfrac{1}{6}$

6. $3\dfrac{5}{6}$ $7\dfrac{1}{2}$ $8\dfrac{7}{8}$ $7\dfrac{5}{6}$

    $-\ 1\dfrac{1}{2}$ $-\ 5\dfrac{1}{3}$ $-\ 6\dfrac{1}{4}$ $-\ 1\dfrac{1}{3}$

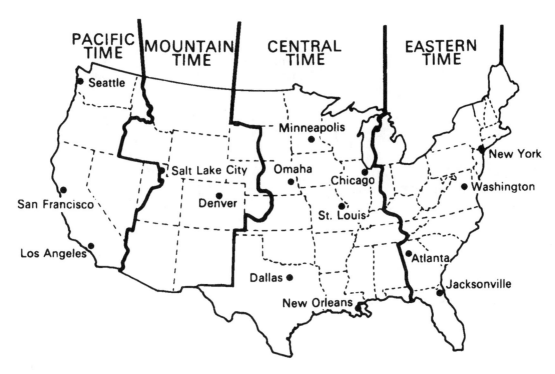

Show the times.

Multiply.

1.

41	63	82	78	42
× 25	× 15	× 21	× 35	× 27

Divide.

2. 5)276   4)183   2)199   3)171   4)134

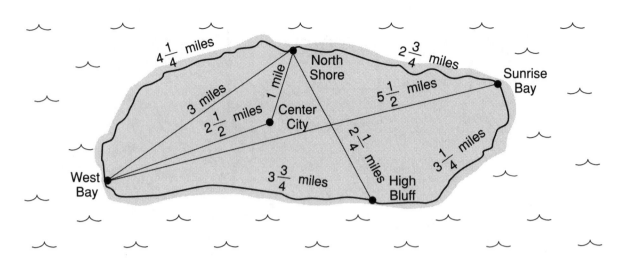

How far ?

1. Start at Sunrise Bay.
Go to North Shore and
on to West Bay on the
coast road. How far ? _____ miles

2. How far from West Bay to
North Shore on the coast road ? _____
How far on the
straight road ? _____
How much longer would it
be on the coast road ? _____

3. Go from West Bay to
High Bluff on the coast
road. Now go on to
Sunrise Bay. How far ? _____

4. Go from High Bluff
to North Shore and
back on the straight
road. How far ? _____

5. Go from Sunrise Bay to
West Bay and back on the
straight road. How far ? _____

Add. Simplify your answers.

1.
$$\frac{2}{5}$$
$$+\frac{3}{5}$$
_____

$$\frac{3}{8}$$
$$+\frac{3}{8}$$
_____

$$\frac{2}{6}$$
$$+\frac{3}{6}$$
_____

$$\frac{3}{7}$$
$$+\frac{3}{7}$$
_____

2.
$$\frac{2}{3}$$
$$+\frac{1}{6}$$
_____

$$\frac{5}{6}$$
$$+\frac{1}{12}$$
_____

$$\frac{4}{5}$$
$$+\frac{2}{3}$$
_____

$$\frac{1}{4}$$
$$+\frac{1}{3}$$
_____

3.
$$2\frac{1}{5}$$
$$+1\frac{3}{10}$$
_____

$$3\frac{1}{6}$$
$$+2\frac{2}{3}$$
_____

$$7\frac{1}{2}$$
$$+4\frac{3}{10}$$
_____

$$3\frac{1}{4}$$
$$+2\frac{5}{8}$$
_____

Subtract. Simplify your answers.

4.
$$\frac{7}{8}$$
$$-\frac{2}{8}$$
_____

$$\frac{5}{6}$$
$$-\frac{1}{6}$$
_____

$$\frac{2}{7}$$
$$-\frac{1}{7}$$
_____

$$\frac{7}{9}$$
$$-\frac{1}{9}$$
_____

5.
$$\frac{7}{9}$$
$$-\frac{1}{3}$$
_____

$$\frac{5}{6}$$
$$-\frac{1}{2}$$
_____

$$6\frac{3}{4}$$
$$-2\frac{1}{2}$$
_____

$$7\frac{2}{3}$$
$$-1\frac{1}{6}$$
_____

Measure these line segments. Write the measurement on the line segment.

1. ⌐————————⌐   ⌐—————————⌐   ⌐——————————————⌐

2. ⌐—————⌐   ⌐—————————————————————⌐

3. ⌐————————————————————⌐   ⌐———⌐

Divide.

4. $2\overline{)162}$ $5\overline{)371}$ $4\overline{)248}$ $3\overline{)197}$

5. $3\overline{)251}$ $4\overline{)399}$ $5\overline{)274}$ $6\overline{)378}$

Multiply.

6.
$$\begin{array}{r} 473 \\ \times\ \ \ 3 \\ \hline \end{array}\qquad \begin{array}{r} 823 \\ \times\ \ \ 7 \\ \hline \end{array}\qquad \begin{array}{r} 641 \\ \times\ \ \ 8 \\ \hline \end{array}\qquad \begin{array}{r} 296 \\ \times\ \ \ 5 \\ \hline \end{array}\qquad \begin{array}{r} 372 \\ \times\ \ \ 4 \\ \hline \end{array}$$

7.
$$\begin{array}{r} 28 \\ \times\ 37 \\ \hline \end{array}\qquad \begin{array}{r} 64 \\ \times\ 28 \\ \hline \end{array}\qquad \begin{array}{r} 93 \\ \times\ 26 \\ \hline \end{array}\qquad \begin{array}{r} 82 \\ \times\ 27 \\ \hline \end{array}\qquad \begin{array}{r} 46 \\ \times\ 13 \\ \hline \end{array}$$

Draw line segments that measure:

$2\frac{1}{2}$"

$3\frac{1}{4}$"

$\frac{1}{4}$"

$4\frac{1}{4}$"

$\frac{1}{2}$"

Multiply.

$$\boxed{\frac{1}{2} \times \frac{1}{4} = \frac{1}{8}}$$

1. $\frac{1}{2} \times \frac{1}{2} =$ —— $\frac{2}{3} \times \frac{1}{3} =$ ——

2. $\frac{1}{2} \times \frac{1}{3} =$ —— $\frac{1}{3} \times \frac{1}{4} =$ —— $\frac{1}{2} \times \frac{3}{8} =$ ——

3. $\frac{1}{5} \times \frac{1}{4} =$ —— $\frac{2}{5} \times \frac{1}{2} =$ —— $\frac{2}{3} \times \frac{1}{2} =$ ——

4.
$$\begin{array}{r} 47 \\ \times\ 15 \\ \hline \end{array}$$
$$\begin{array}{r} 76 \\ \times\ 23 \\ \hline \end{array}$$
$$\begin{array}{r} 84 \\ \times\ 21 \\ \hline \end{array}$$
$$\begin{array}{r} 60 \\ \times\ 73 \\ \hline \end{array}$$

dividing fractions

Measure these line segments. Write the measurement on the line segment.

1. |_____| |_____|

2. |__| |_____| |_____| |____|

3. |_____| |_____|

Multiply.

4. $\dfrac{1}{4} \times \dfrac{1}{3} = $ ___ $\dfrac{1}{3} \times \dfrac{2}{3} = $ ___ $\dfrac{3}{5} \times \dfrac{1}{2} = $ ___

5. $\dfrac{2}{3} \times \dfrac{1}{5} = $ ___ $\dfrac{2}{3} \times \dfrac{1}{4} = $ ___ $\dfrac{1}{2} \times \dfrac{1}{6} = $ ___

6. $\dfrac{1}{3} \times \dfrac{1}{2} = $ ___ $\dfrac{3}{4} \times \dfrac{1}{2} = $ ___ $\dfrac{2}{3} \times \dfrac{1}{2} = $ ___

Divide.

$$\dfrac{1}{2} \div \dfrac{1}{2} = \text{___}$$
$$\dfrac{1}{2} \times \dfrac{2}{1} = \dfrac{2}{2} = 1$$

7. $\dfrac{2}{3} \div 2 = $ ___ $\dfrac{3}{4} \div \dfrac{1}{4} = $ ___

$\dfrac{2}{3} \times \dfrac{1}{2} = $ ___ $\dfrac{3}{4} \times \dfrac{4}{1} = $ ___

8. $\dfrac{3}{8} \div \dfrac{1}{2} = $ ___ $\dfrac{2}{3} \div \dfrac{1}{2} = $ ___ $\dfrac{2}{3} \div \dfrac{2}{3} = $ ___

$\dfrac{3}{8} \times \dfrac{2}{1} = $ ___ $\dfrac{2}{3} \times \dfrac{2}{1} = $ ___ $\dfrac{2}{3} \times \dfrac{3}{2} = $ ___

9. $\dfrac{1}{3} \div \dfrac{1}{2} = $ ___ $\dfrac{1}{2} \div \dfrac{1}{3} = $ ___ $\dfrac{1}{4} \div \dfrac{1}{2} = $ ___

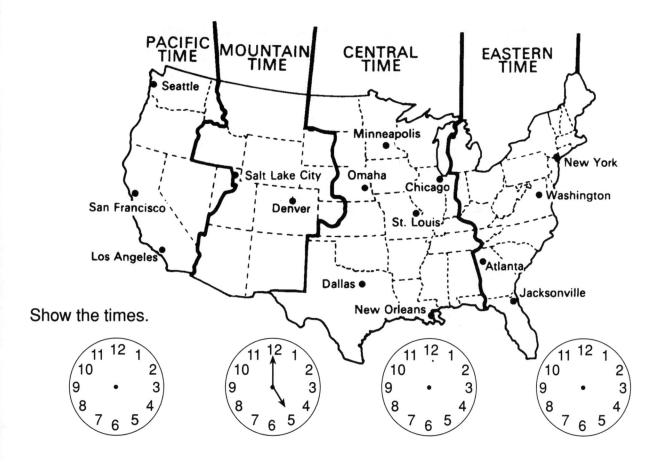

Show the times.

Multiply.

$$\begin{array}{c} 1 \\ \dfrac{\cancel{3}}{4} \times \dfrac{1}{\cancel{6}_{2}} = \dfrac{1}{8} \end{array}$$

1. $\dfrac{1}{2} \times \dfrac{1}{4} = $ —— $\qquad \dfrac{3}{8} \times 8 = $ ——

2. $\dfrac{1}{6} \times \dfrac{3}{5} = $ —— $\qquad \dfrac{2}{3} \times \dfrac{1}{6} = $ ——

Divide.

$$3 \div \dfrac{1}{8} = \text{——}$$

$$\dfrac{3}{1} \div \dfrac{8}{1} = \dfrac{24}{1} = 24$$

3. $9 \div \dfrac{1}{3} = $ —— $\qquad 2 \div \dfrac{1}{3} = $ ——

$\quad \dfrac{9}{1} \times \dfrac{3}{1} = $ —— $\qquad \dfrac{2}{1} \times \dfrac{3}{1} = $ ——

4. $\dfrac{1}{4} \div \dfrac{1}{2} = $ —— $\qquad \dfrac{3}{4} \div \dfrac{3}{5} = $ ——

Measure these line segments. Write the measurements on the line segment.

1. ⌐————————————⌐   ⌐————————⌐   ⌐——⌐

2. ⌐————————————⌐   ⌐————⌐

Add.

3.
$\dfrac{2}{3}$ $\dfrac{2}{3}$ $\dfrac{3}{4}$ $1\dfrac{2}{5}$ $2\dfrac{1}{2}$

$+\dfrac{1}{3}$ $+\dfrac{1}{2}$ $+\dfrac{1}{8}$ $+\,3\dfrac{2}{5}$ $+\,3\dfrac{1}{4}$

Subtract.

4.
$\dfrac{8}{9}$ $\dfrac{3}{5}$ $\dfrac{2}{3}$ $2\dfrac{1}{2}$ $3\dfrac{2}{3}$

$-\dfrac{1}{9}$ $-\dfrac{1}{5}$ $-\dfrac{1}{6}$ $-\,1\dfrac{1}{4}$ $-\,1\dfrac{1}{6}$

Multiply or divide.

$\dfrac{1}{3} \times 6 = $ ———
$\dfrac{1}{\cancel{3}_1} \times \dfrac{\cancel{6}^2}{1} = \dfrac{2}{1} = 2$

5. $\dfrac{1}{3} \times \dfrac{1}{2} = $ —— $\dfrac{2}{5} \times 10 = $ ——

6. $\dfrac{1}{8} \times \dfrac{1}{2} = $ —— $\dfrac{1}{2} \times 4 = $ ——

$\dfrac{3}{5} \div \dfrac{1}{2} = $ ——
$\dfrac{3}{5} \div \dfrac{2}{1} = \dfrac{6}{5} = 1\dfrac{1}{5}$

7. $2 \div \dfrac{2}{3} = $ —— $\dfrac{3}{4} \div \dfrac{1}{2} = $ ——

You bought sneakers for $42.00.
You paid Sam's shoes with this check.

		001
		71-00 / 000

Date _____

PAY TO THE ORDER OF *Sam's Shoes* _____ $ *42.00*

forty-two and ⁿᵒ⁄₁₀₀ _____ DOLLARS

PIONEER BANK
WESTVIEW, FL.

Your name

Memo _____

⑆⑆6⑈34⑈9023⑈   647 009⑈⑆

Add the prices. Pay Smith's Video store with a check.

$7.00
2.58
+ 1.29

		002
		71-00 / 000

Date _____

PAY TO THE ORDER OF _____ $ _____

_____ DOLLARS

PIONEER BANK
WESTVIEW, FL.

Memo _____

⑆⑆6⑈34⑈9023⑈   647 009⑈⑆

Add.

1.

$3.84	$62.34	$20.39	$4.93	$6.38
+ 1.73	+ 3.86	+ 14.70	+ 1.72	+ 1.72

2. Add these purchases. Write a check to Joe's Pet Shop.

$3.27
8.36
1.93
+ .27

```
                                                          003
                                                          71-00
                                 Date _____       000

PAY TO THE                                          $
ORDER OF  _____

_____ DOLLARS

PIONEER BANK
WESTVIEW, FL.

Memo _____      _____

⑆6134⑈9023⑆   647  009⑈
```

Subtract.

3.

$7.21	$6.31	$2.37	$6.09	$2.34
− .30	− 1.80	− 1.28	− 1.73	− 1.12

4. You have $149.00 in the checkbook. You write a check to Dr. Jones for $27.00. How much is left in the account ? _____

1.

$$32 \times 51$$ $$37 \times 83$$ $$64 \times 25$$ $$93 \times 38$$ $$61 \times 27$$

2.

$$\$3.84 \times 3$$ $$\$7.21 \times 2$$ $$\$3.87 \times 5$$ $$\$6.09 \times 6$$ $$\$2.93 \times 5$$

3. Sandwiches cost $3.15.
We bought 4 sandwiches.
How much did they cost ? _____

4. Milkshakes cost $2.59.
We bought 3 of them.
How much did they cost ? _____

Divide.

5. $3\overline{)214}$ $2\overline{)179}$ $5\overline{)372}$

6. $4\overline{)\$3.80}$ $5\overline{)\$3.25}$ $2\overline{)\$1.92}$

7. A giant sandwich cost $2.88.
Two boys split the sandwich
and shared the cost. How
much did each boy pay ? _____

You have:	You buy:	Total Cost	Change
75¢	51¢ 10¢	5 1 ¢ + 1 0 ¢ 6 1 ¢	7 5 ¢ − 6 1 ¢ 1 4 ¢
1. 89¢	62¢ 21¢		
2. $1.25	$.50 $.65		
3. $2.75	$.82 $1.25		
4. $5.00	1.25 $.79		
5. $5.00	$.18 $1.35 $.79		
6. $6.00	$3.92 $1.25 $.17		
7. $3.00	$2.25 $.57		

	Cost	You buy this many	Total cost
	$.28	3	$$\begin{array}{r} 2 \\ \$.28 \\ \times\ \ \ 3 \\ \hline .84 \end{array}$$
1.	$1.27	5	
2.	$.45	12	
3.	$13.42	4	
4.	$26.23	7	
5.	$14.95	6	
6.	$.58	25	
7.	$.93	15	
8.	$1.27	10	

1. A bag of pretzels costs $2.38.
   Two girls share the cost.
   How much for each girl ? _____

2. Don earned $56.00
   for walking a dog
   for four weeks.
   How much did
   he earn each week ? _____

3. A six-pack of water
   costs $4.74. How much
   much for each bottle ? _____

Divide.

4. $2\overline{)\$1.82}$ $3\overline{)\$1.44}$ $5\overline{)\$3.95}$

5. $7\overline{)\$3.64}$ $8\overline{)\$6.56}$ $9\overline{)\$6.48}$

Add.

1. $3.80 $4.27 $8.09 $14.64 $12.97
   + 1.19 + .16 + 2.47 + 23.28 + 2.83

Subtract.

2. $1.47 $2.00 $14.34 $13.42 $20.00
   - .29 - 1.83 - 7.21 - 7.91 - 17.43

Multiply.

3. 42 $16.42 $1.39 $.46 72
   × 28 × 5 × 7 × 23 × 89

Divide.

4. 2)82 2)31 4)53 3)91

5. 6)$2.94 3)$2.64 8)$4.48 5)$6.65

1. A straight **line** does not change direction. It goes on forever in both directions. Circle the **line**:

2. A **line segment** is part of a line. It is straight. It has two endpoints. Circle the **line segment**:

3. A **ray** is part of a line. It is straight. It has one endpoint. It goes on forever in one direction. Circle the **ray**:

4. **Parallel lines** are lines that are always the same distance apart. Circle the **parallel lines**:

5. **Perpendicular lines** are lines that meet to form a square corner. Circle the **perpendicular lines**:

6. **Horizontal lines** are lines that go straight across. Circle the **horizontal line**:

7. **Vertical lines** are lines that go straight up and down. Circle the **vertical line**:

8. **Intersecting lines** are lines that meet or cross each other. Circle the **intersecting lines**:

9. It is easier to describe lines if they are labeled. Circle the diagram that shows **line AB is parallel to line CD.**

| A ———— B | A | A ———— B |
| D ———— F | B └— C | C ———— D |

10. Circle the diagram that shows **line FG is perpendicular to line GH.**

A ———— B	F	F ———— G
F ———— G	┼—H	E ———— H
	G	

| line | line segment | parallel lines |
| ray | intersecting lines | perpendicular lines |

Use the word bank to label these diagrams.

_____

_____

_____

_____

_____

_____

Add.

1.
$$\begin{array}{r} \$38.24 \\ +\ 10.23 \\ \hline \end{array} \qquad \begin{array}{r} \$26.92 \\ +\ 31.51 \\ \hline \end{array} \qquad \begin{array}{r} \$36.24 \\ +\ 28.71 \\ \hline \end{array} \qquad \begin{array}{r} \$30.03 \\ +\ 27.32 \\ \hline \end{array} \qquad \begin{array}{r} \$68.93 \\ +\ 1.99 \\ \hline \end{array}$$

2.
$$\begin{array}{r} \$1.48 \\ +\ 2.83 \\ \hline \end{array} \qquad \begin{array}{r} \$1.97 \\ +\ 3.84 \\ \hline \end{array} \qquad \begin{array}{r} \$37.42 \\ +\ 14.63 \\ \hline \end{array} \qquad \begin{array}{r} \$40.93 \\ +\ 27.76 \\ \hline \end{array} \qquad \begin{array}{r} \$82.14 \\ +\ 27.93 \\ \hline \end{array}$$

Subtract.

3.
$$\begin{array}{r} \$1.43 \\ -\ .27 \\ \hline \end{array} \qquad \begin{array}{r} \$2.96 \\ -\ 1.83 \\ \hline \end{array} \qquad \begin{array}{r} \$14.82 \\ -\ 6.91 \\ \hline \end{array} \qquad \begin{array}{r} \$32.00 \\ -\ 17.05 \\ \hline \end{array} \qquad \begin{array}{r} \$61.38 \\ -\ 27.42 \\ \hline \end{array}$$

4.
$$\begin{array}{r} \$18.93 \\ -\ 9.94 \\ \hline \end{array} \qquad \begin{array}{r} \$6.37 \\ -\ 1.99 \\ \hline \end{array} \qquad \begin{array}{r} \$37.42 \\ -\ 25.21 \\ \hline \end{array} \qquad \begin{array}{r} \$64.36 \\ -\ 28.93 \\ \hline \end{array} \qquad \begin{array}{r} \$27.94 \\ -\ 12.36 \\ \hline \end{array}$$

Straight lines may be used to make designs.

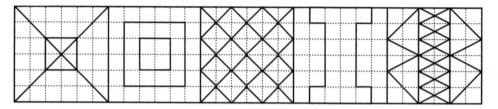

Copy the designs.

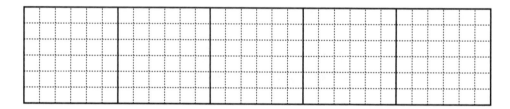

Make your own designs.

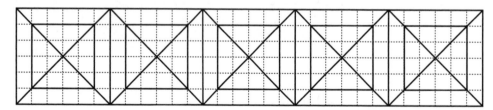

Lines can make repeating patterns.

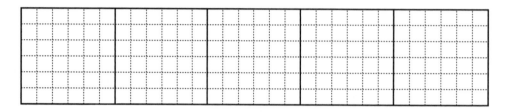

Copy the patterns.

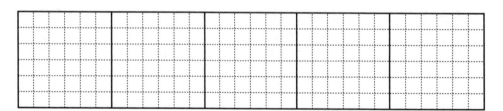

circle	◯	octagon	⬡	cube	
square	▢	pentagon	⬠	pyramid	
rectangle	▭	oval	⬭	sphere	
triangle	△	diamond	◇	cylinder	
rectangular prism		cone	▽		

Use the word bank to identify shapes.

1.

_____   _____   _____   _____

2.

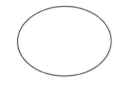   

_____   _____   _____   _____

3.

         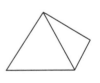

_____   _____   _____   _____

4.

   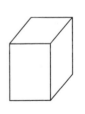

_____   _____   _____   _____

Use the word bank to identify shapes.

| circle | square | oval |
| triangle | rectangle | diamond |

1.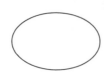

_____   _____   _____   _____

2.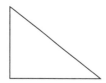

_____   _____   _____   _____

Use the word bank to identify solid figures.

| cube | sphere | cone |
| pyramid | cyclinder | rectangular prism |

3.

_____   _____   _____   _____

4.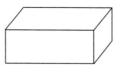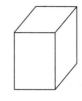

_____   _____   _____   _____

5.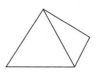

_____   _____   _____   _____

Draw these :

circle   square   triangle   rectangle

oval   diamond   octogan   cube

Draw these:

line   vertical line

line segment   parallel lines

ray   perpendicular lines

horizontal line   intersecting lines

Add.

1.	323	629	403	623	387
	+ 174	+ 283	+ 709	+ 794	+ 423

Subtract.

2.	382	609	300	252	821
	− 173	− 593	− 176	− 163	− 746

Use the word bank. Name the figure the object looks like.

circle	rectangle	oval	cube
square	triangle	pyramid	sphere

1.

_____   _____   _____   _____

2.

_____   _____   _____   _____

Multiply.

3.
$$\begin{array}{r} 34 \\ \times\,25 \\ \hline \end{array} \qquad \begin{array}{r} 72 \\ \times\,36 \\ \hline \end{array} \qquad \begin{array}{r} 40 \\ \times\,29 \\ \hline \end{array} \qquad \begin{array}{r} \$9.37 \\ \times\qquad 7 \\ \hline \end{array} \qquad \begin{array}{r} \$4.29 \\ \times\qquad 6 \\ \hline \end{array}$$

Divide.

4. $6\overline{)49}$ $\qquad$ $3\overline{)123}$ $\qquad$ $2\overline{)47}$ $\qquad$ $5\overline{)37}$ $\qquad$ $6\overline{)69}$

5. $2\overline{)\$3.20}$ $\qquad$ $4\overline{)\$2.16}$ $\qquad$ $5\overline{)\$3.95}$ $\qquad$ $7\overline{)\$4.34}$ $\qquad$ $9\overline{)\$3.69}$

Perimeter is the distance around an object. Add the length of each side to find the perimeter.

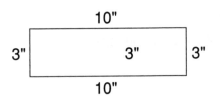

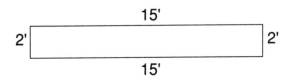

10"
3"
10"
3"

26" perimeter

1.

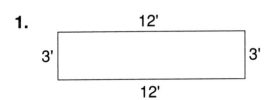

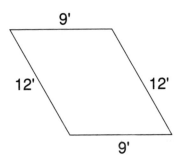

2.

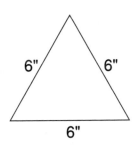

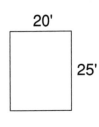

3.

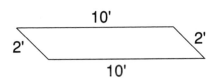

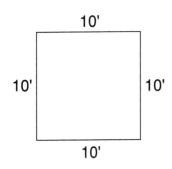

4.

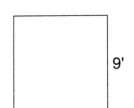

Find the perimeter of each figure.
Use the words in the word bank to name shapes.

| triangle | rectangle | square | pentagon | diamond |

1.

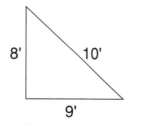

name _____triangle_____

perimeter _____

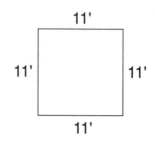

name _____

perimeter _____

2.

name _____

perimeter _____

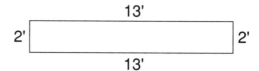

name _____

perimeter _____

3.

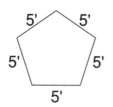

name _____

perimeter _____

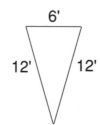

name _____

perimeter _____

4.

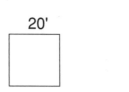

name _____

perimeter _____

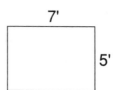

name _____

perimeter _____

Area is the number
of square units needed
to cover a flat space.

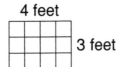

4 feet

3 feet

$A = L \times W$

4 feet
$\times$ 3 feet
———————
12 square feet

Area = Length × Width

1. 5 feet

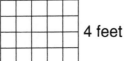

4 feet

5 inches
$\times$ 4 inches
—————

6 ft

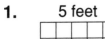

5 ft

2.

2 ft

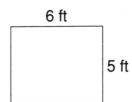

12 ft

8 ft

5 ft

3. 15 ft

3 ft

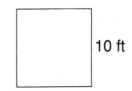

10 ft

4. 10 ft

4 ft

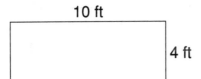

9 miles

4 miles

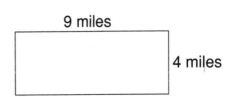

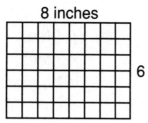

8 inches

6 inches

Perimeter	Area
8 inches	
6 inches	
8 inches	8 inches
+ 6 inches	× 6 inches
26 inches	48 square inches

Find the perimeters.

1.

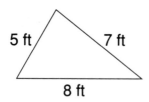

5 ft   7 ft

8 ft

8'

8'

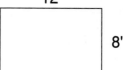

12'

8'

———

———

———

2.

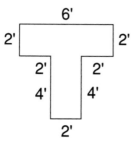

6'

2'   2'

2'   2'

4'   4'

2'

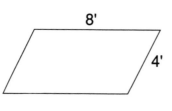

8'

4'

4'

6'

———

———

———

Find the areas.

3.

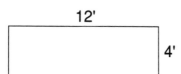

12'

4'

9'

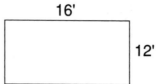

30'

6'

———

———

———

4.

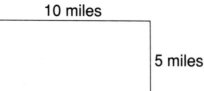

10 miles

5 miles

16'

12'

12'

———

———

Triangles have 3 sides
and 3 angles.

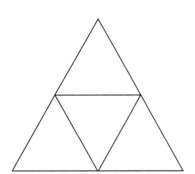 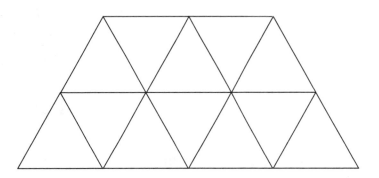

How many s do you count ? _____

Find s that have the same size and shape. Draw lines to match them.

Color matching s the same color.

An angle is measured in degrees.
A protractor is used to measure angles.
This is a protractor:

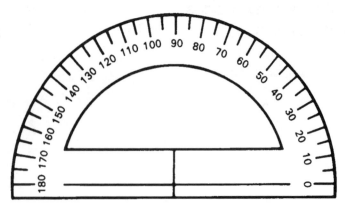

This is how you measure an angle.

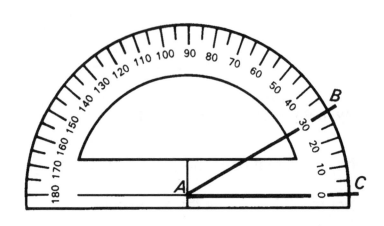

How many degrees are there in
each of these angles ?

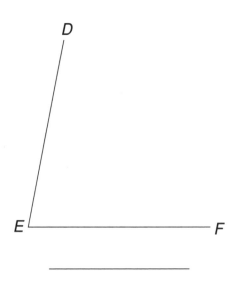

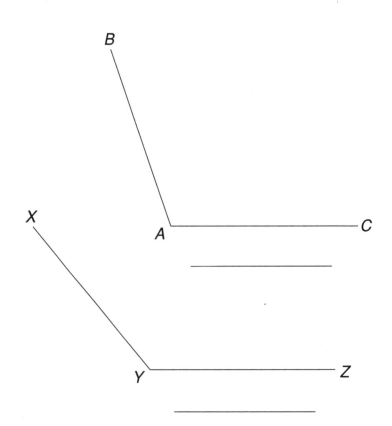

Right angles have perpendicular lines that make a square corner.

Acute angles measure less than a right angle.

Obtuse angles measure greater than a right angle.

 90°   90°

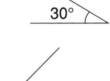
30°

45°

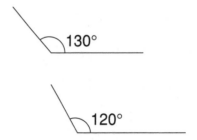
130°

120°

Measure these angles. Write the measurement inside the angles.
Label each angle as **right** angle, **acute** angle or **obtuse** angle.

1.

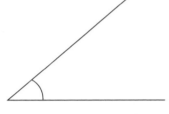

_____ angle   _____ angle   _____ angle

2.

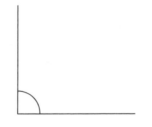
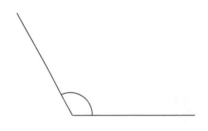

_____ angle   _____ angle   _____ angle

3.

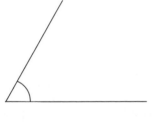
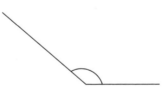

_____ angle   _____ angle   _____ angle

Draw angles that measure:

90°   40°   100°

120°   20°   70°

Add.

1.
23	13	68	74	19
56	15	50	68	38
84	25	91	20	27
+ 31	+ 93	+ 37	+ 57	+ 34

2.
427	209	463	3042	3947
+ 386	+ 387	+ 294	+ 7069	+ 7821

Subtract.

3.
84	63	842	347	900
− 24	− 27	− 171	− 258	− 243

Check off steps as you complete each one.

☐ Measure the angles of the large .

☐ Trace the 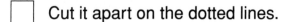 onto blank paper.

☐ Cut out the triangle.

☐ Cut it apart on the dotted lines.

☐ Place the corner angles next to each other.

☐ The corner angles form a straight line.

☐ Add the angle measurements together.

☐ A straight line and 3 angles of a 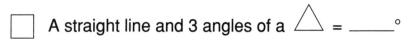 = _____°

Try the same thing with other 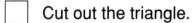s.

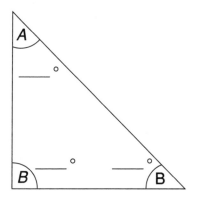

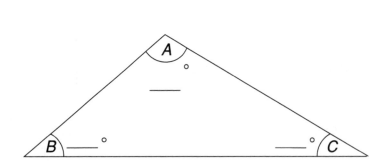

A compass is used to draw circles.

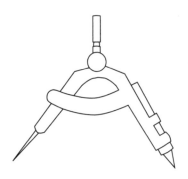

center

diameter

radius

circumference

Put labels on each circle.
Use these words:
  circumference
  center
  radius
  diameter

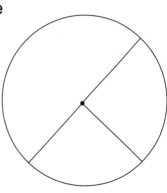

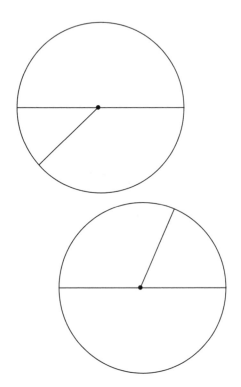

The radius is $\frac{1}{2}$ the diameter. Find the radius.

1. (10) d = 10   (12) d = 12   (8) d = 8

r = ____   r = ____   r = ____

2. The diameter is twice the radius. Find the diameter.

(14) r = 14   (7) r = 7   (6) r = 6

d = ____   d = ____   d = ____

Use a compass and a ruler.

Draw a circle.

Label the center.

Draw the diameter.

Label the diameter.

Measure the diameter. _____

Draw the radius.

Label the radius.

Measure the radius. _____

Label the circumference.

Multiply.

1.
$$\begin{array}{r} 21 \\ \times\ 84 \\ \hline \end{array}$$
$$\begin{array}{r} 68 \\ \times\ 23 \\ \hline \end{array}$$
$$\begin{array}{r} 97 \\ \times\ 85 \\ \hline \end{array}$$
$$\begin{array}{r} 907 \\ \times\ \ \ 6 \\ \hline \end{array}$$
$$\begin{array}{r} 829 \\ \times\ \ \ 5 \\ \hline \end{array}$$

Divide.

2. $6\overline{)52}$ $5\overline{)46}$ $2\overline{)41}$ $3\overline{)37}$ $5\overline{)56}$

3. $2\overline{)47}$ $5\overline{)154}$ $6\overline{)238}$ $7\overline{)442}$ $9\overline{)800}$

line	line segment	vertical line
parallel lines	rectangle	circle
perpendicular lines	cube	cylinder
triangle	horizontal line	pyramid

1.

_____   _____   _____

2.

_____   _____   _____

3.

_____   _____   _____

Find the perimeter.

4.

 23'   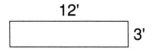 3' 12'   17'

_____   _____   _____

5.

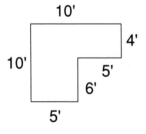   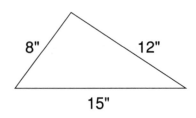

_____   _____

Find the areas.

1. 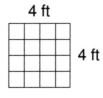 4 ft

4 ft

5 ft 2 ft

3 ft 4 ft

2. 8 ft

3 ft

6 ft

6 ft

10 ft

5 ft

Measure these angles. Write the measurements inside the angles.

3.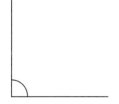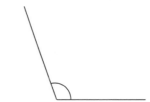

Use the word bank. Write the names.

acute angle	right angle	compass
circle	obtuse angle	protractor

4.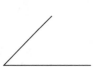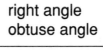

5.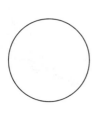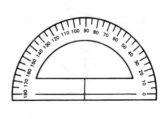

Color these designs.

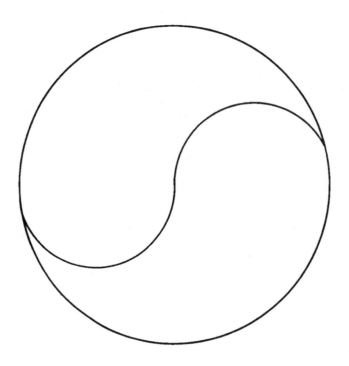

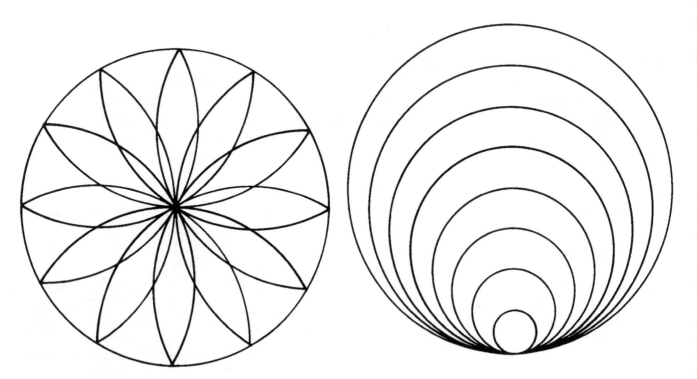

Try to make some designs of your own.

1. Write the temperature.

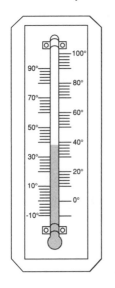

_____

2. Count the money.

_____¢

Add.

3.
```
   406        297        642        392        406
 + 133      + 104      + 258      + 138      + 299
```

4.
```
    38        127      $42.83     $27.93     $62.84
    27         93      +  2.98    + 14.13    +  7.99
    65        142
 +  36       + 63
```

Subtract.

5.
```
    67         83         40         63         46
 -  25       - 12       - 15       - 29       - 17
```

6.
```
   323        624        326      $34.27     $16.23
 - 161       - 536      - 183     - 21.38    -  5.38
```

1. _____   _____   _____   _____   _____

Multiply.

2.
$$\begin{array}{r} 8 \\ \times\,7 \\ \hline \end{array} \qquad \begin{array}{r} 9 \\ \times\,6 \\ \hline \end{array} \qquad \begin{array}{r} 3 \\ \times\,7 \\ \hline \end{array} \qquad \begin{array}{r} 4 \\ \times\,8 \\ \hline \end{array} \qquad \begin{array}{r} 9 \\ \times\,7 \\ \hline \end{array} \qquad \begin{array}{r} 8 \\ \times\,9 \\ \hline \end{array} \qquad \begin{array}{r} 6 \\ \times\,8 \\ \hline \end{array} \qquad \begin{array}{r} 7 \\ \times\,6 \\ \hline \end{array} \qquad \begin{array}{r} 7 \\ \times\,7 \\ \hline \end{array}$$

3.
$$\begin{array}{r} 43 \\ \times\,2 \\ \hline \end{array} \qquad \begin{array}{r} 46 \\ \times\,5 \\ \hline \end{array} \qquad \begin{array}{r} 183 \\ \times\,\,\,7 \\ \hline \end{array} \qquad \begin{array}{r} 26 \\ \times\,15 \\ \hline \end{array} \qquad \begin{array}{r} 94 \\ \times\,27 \\ \hline \end{array} \qquad \begin{array}{r} 86 \\ \times\,35 \\ \hline \end{array}$$

Divide.

4. $8\overline{)64}$ $\qquad$ $9\overline{)72}$ $\qquad$ $8\overline{)73}$ $\qquad$ $6\overline{)49}$ $\qquad$ $5\overline{)37}$

5. $3\overline{)207}$ $\qquad$ $4\overline{)63}$ $\qquad$ $2\overline{)79}$ $\qquad$ $5\overline{)375}$ $\qquad$ $6\overline{)400}$

1. Measure these line segments.

Add or subtract fractions.

2.

$$\begin{array}{r} \frac{4}{9} \\ + \frac{7}{9} \\ \hline \end{array}$$

$$\begin{array}{r} \frac{3}{5} \\ + \frac{3}{5} \\ \hline \end{array}$$

$$\begin{array}{r} \frac{1}{2} \\ + \frac{2}{3} \\ \hline \end{array}$$

$$\begin{array}{r} \frac{3}{4} \\ + \frac{1}{8} \\ \hline \end{array}$$

$$\begin{array}{r} 7\frac{1}{2} \\ + 3\frac{1}{2} \\ \hline \end{array}$$

3.

$$\begin{array}{r} \frac{7}{8} \\ - \frac{3}{8} \\ \hline \end{array}$$

$$\begin{array}{r} \frac{2}{3} \\ - \frac{1}{2} \\ \hline \end{array}$$

$$\begin{array}{r} 4\frac{5}{8} \\ - 2\frac{3}{8} \\ \hline \end{array}$$

$$\begin{array}{r} 5\frac{3}{4} \\ - 1\frac{1}{4} \\ \hline \end{array}$$

$$\begin{array}{r} 6\frac{2}{3} \\ - 1\frac{1}{6} \\ \hline \end{array}$$

Multiply or divide. Simplify your answers.

4. $\frac{1}{2} \times \frac{1}{4} =$ $\frac{2}{3} \times 9 =$ $7 \times \frac{1}{2} =$

5. $8 \div \frac{1}{2} =$ $6 \div \frac{1}{3} =$ $\frac{1}{2} \div \frac{1}{2} =$

Use data from the chart to complete the line graph.

Justin's Scores on
Multiplication Fact Tests

Week	Score
1	65
2	65
3	67
4	75
5	82
6	90
7	96
8	100

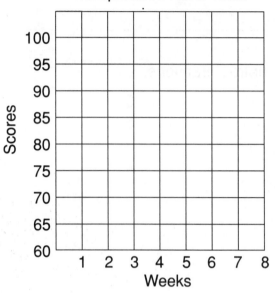

Multiplication Facts Tests

MAGIC STAR

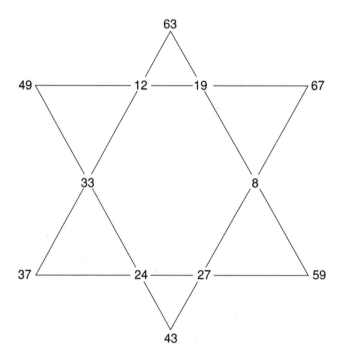

Find the sums of the three numbers around each star point.

_____ _____ _____ _____ _____ _____

Find the sums of the rows of four numbers. _____ _____

Find the sums of the numbers around each large triangle. _____ _____